Global Warming and Climate Change Demystified

Demystified Series

Accounting Demystified
Advanced Calculus Demystified
Advanced Physics Demystified
Advanced Statistics Demystified
Algebra Demystified
Alternative Energy Demystified
Anatomy Demystified
asp.net 2.0 Demystified
Astronomy Demystified
Audio Demystified
Biology Demystified
Biotechnology Demystified
Business Calculus Demystified
Business Math Demystified
Business Statistics Demystified
C++ Demystified
Calculus Demystified
Chemistry Demystified
Circuit Analysis Demystified
College Algebra Demystified
Corporate Finance Demystified
Data Structures Demystified
Databases Demystified
Diabetes Demystified
Differential Equations Demystified
Digital Electronics Demystified
Earth Science Demystified
Electricity Demystified
Electronics Demystified
Engineering Statistics Demystified
Environmental Science Demystified
Everyday Math Demystified
Fertility Demystified
Financial Planning Demystified
Forensics Demystified
French Demystified
Genetics Demystified
Geometry Demystified
German Demystified
Global Warming and Climate Change Demystified
Hedge Funds Demystified
Home Networking Demystified
Investing Demystified
Italian Demystified
Java Demystified
JavaScript Demystified
Lean Six Sigma Demystified
Linear Algebra Demystified
Macroeconomics Demystified
Management Accounting Demystified
Math Proofs Demystified
Math Word Problems Demystified
MATLAB® Demystified
Medical Billing and Coding Demystified
Medical Terminology Demystified
Meteorology Demystified
Microbiology Demystified
Microeconomics Demystified
Nanotechnology Demystified
Nurse Management Demystified
OOP Demystified
Options Demystified
Organic Chemistry Demystified
Personal Computing Demystified
Pharmacology Demystified
Physics Demystified
Physiology Demystified
Pre-Algebra Demystified
Precalculus Demystified
Probability Demystified
Project Management Demystified
Psychology Demystified
Quality Management Demystified
Quantum Mechanics Demystified
Real Estate Math Demystified
Relativity Demystified
Robotics Demystified
Sales Management Demystified
Signals and Systems Demystified
Six Sigma Demystified
Spanish Demystified
SQL Demystified
Statics and Dynamics Demystified
Statistics Demystified
Technical Analysis Demystified
Technical Math Demystified
Trigonometry Demystified
UML Demystified
Visual Basic 2005 Demystified
Visual C# 2005 Demystified
Vitamins and Minerals Demystified
XML Demystified

Global Warming and Climate Change Demystified

Jerry Silver

New York Chicago San Francisco Lisbon London
Madrid Mexico City Milan New Delhi San Juan
Seoul Singapore Sydney Toronto

The McGraw-Hill Companies

Library of Congress Cataloging-in-Publication Data

Silver, Jerry.
Global warming and climate change demystified / Jerry Silver.—1st ed.
p. cm.
ISBN 978-0-07-150240-5 (alk. paper)
1. Global warming—Popular works. 2. Climatic changes—Popular works. I. Title.
QC981.8.G56S548 2008
551.6—dc22 2007046678

1 2 3 4 5 6 7 8 9 0 DOC/DOC 0 1 4 3 2 1 0 9 8

ISBN 978-0-07-150240-5
MHID 0-07-150240-8

Sponsoring Editor
Judy Bass

Production Supervisor
Pamela A. Pelton

Editing Supervisor
Stephen M. Smith

Project Manager
Harleen Chopra, International Typesetting and Composition

Copy Editor
James K. Madru

Proofreader
Honey Paul

Indexer
Broccoli Information Management

Art Director, Cover
Jeff Weeks

Composition
International Typesetting and Composition

Printed and bound by RR Donnelley.

This book is dedicated to my wife, Joanie, who is, and has always been, ready to take the next positive step, and to Ally and Danny, whose future grandchildren deserve to inherit a stable planet.

ABOUT THE AUTHOR

Jerry Silver has worked for nearly two decades supporting the development and manufacturing of solar cells. He was a design engineer for solar arrays used to provide power for commercial and NASA satellites. He holds a B.S. in Engineering Physics from Cornell University and a M.S. in Physics from the University of Massachusetts. Mr. Silver currently teaches science in New Jersey.

CONTENTS AT A GLANCE

CONTENTS

ACKNOWLEDGMENTS

The author would like to thank Judy Winchock, Bob Moshman, the Kaolins, and Joan Silver for their help with the text. Grateful recognition also goes to the many dedicated men and women around the world who have collected and assembled the scientific data that this book is based on.

Global Warming and Climate Change Demystified

CHAPTER 1

Global Perspective—Thinking about the Earth

To fully understand how the earth's climate is changing, it is important to maintain a global perspective. Conclusions about the earth's climate are based on small changes to global averages viewed in the context of the entire history of the earth. Collecting the data is a massive effort requiring the collaboration of many scientists around the world. It is understandable how people looking at parts of this data in isolation, can arrive at different conclusions.

In the time it takes to read a few pages of this book, the temperature outside easily may have changed by more than the total average global temperature increase for the entire past century. The rise and fall of the ocean during any given hour far exceeds the overall average global increase in sea level for an entire decade.

Over the last hundred years, the earth's average temperature has increased by about three quarters of a degree Centigrade (1.3°F). This seemingly minor change may strike some people as inconsequential. However, if global warming adds just another 4°C (7°F) or so to the atmospheric temperature, both Greenland and Antarctica could be well on their way to meltdown. A far more conspicuous confirmation that the earth's climate is changing can be seen around the world in the rapid melting of glaciers and ice sheets that had been around for thousands of years.

It is like the scene from the movie, *Jurassic Park,* where the tour guide notices the very first tremors in a glass of water caused by the approaching Tyrannosaurus Rex. The tremors in the earth's climate may be barely detectable at first, but have the potential to reshape our entire planet. How should we respond to these early warning signs? If the earth were a machine shop in Osaka, or a semiconductor processing line in Palo Alto, it would be shut down for being statistically out of control. The sky is not falling just yet; but there may be a very small window of opportunity between when we detect the early warning signs and when it is too late to respond effectively.

One way to define progress is how the world deals with what it needs to dispose of. The industrialized world has progressed in how it has handled sewage treatment, recycling, and reducing sulfur emissions that cause acid rain. Lead was removed from gasoline. Mercury is being captured from smokestacks. Chemicals that destroy ozone were replaced. The American eagle, once threatened by pesticides, was just taken off the endangered species list in 2007. With each success story, there was no doubt someone who insisted that the solutions were unnecessary and would be too expensive. Greenhouse gases are the world's next challenge. The key questions are: In whose backyard will the solutions be located and how much will it cost? Costs can also be put in perspective and weighed against the prospect of living on a planet where greenhouse gas levels continue to be driven off the charts.

Many of the insights about climate change described in this book are the result of work performed by scientists working with The Intergovernmental Panel on Climate Change (IPCC). The IPCC, along with Al Gore, won the 2007 Nobel Peace Prize. This choice by the Nobel committee reflects the belief that the consequences of drastic climate change could lead to instability around the world with an increased risk of conflict.

This book is organized into three parts. Part One is called "What We Know and How We Know It." The starting point for understanding climate change comes from the basic scientific processes of measurement and observation. Scientists around the world have painstakingly collected data from weather instruments, satellite telemetry, ice cores, and coral sections. This book will explore how those data are collected and analyzed.

Next comes Part Two, which is called "Why Climate Changes." The temperature of the earth is the result of well-understood physical processes. Nature is responsible for many climate changes. However, the carbon dioxide that humans have generated from fossil fuel combustion can be considered is a natural atmospheric component at unnatural levels. Greenhouse gas concentrations have never been this high before, and this book will examine their impact.

Part Three is called "What We Can Expect and What We Can Do." No single country is responsible for the climate changes that are occurring, and no single country can reverse them single-handedly. However, to stabilize the earth's temperature, countries around the world may need to fundamentally rethink how they use energy—a process that, for several reasons, may be long overdue. This section will also address the tools that scientists use to forecast climate changes in the future.

The science fiction writer Robert A. Heinlein said, "Climate is what you expect; weather is what you get." *Global Warming and Climate Change Demystified* is intended to help guide the reader to a better understanding of what to expect from the changes that have begun to take place.

PART ONE

What We Know and How We Know It

CHAPTER 2

Taking the Earth's Temperature

Science typically starts with measurement. This chapter is about the data that scientists around the world use to form judgments about climate change. We will discuss how measurements are made and how they are assembled to characterize the climate of the entire earth. There is more than one type of temperature-measuring instrument in the climate scientist's toolkit. These include direct measurements such as thermometers on the earth's surface, ships and buoys at sea, and balloons released into the atmosphere. Satellites in earth orbit currently are providing a much more comprehensive thermal profile of the earth's surface and atmosphere. Scientists also do detective work to investigate what the earth's climate was like as far back as hundreds of thousands of years ago. We know that global warming is occurring because that is what the data shows. Let's start by taking a look at where the data comes from.

Global Temperature Measurement—The Basics

Of the various ways to measure the earth's temperature, the simplest and most straightforward method is to use a thermometer. Galileo developed the first known thermometer in the early 1600s. His thermometer was crude by today's standards and only narrowed air temperature readings to a fairly broad temperature range. But this was a start. Experimenters in Europe introduced liquid-filled bulb thermometers that were more reliable and accurate. In 1714, Gabriel Fahrenheit invented the first mercury thermometer, and it became the first established method for tracking the changes that the earth's climate was beginning to experience.

There is no single location where we can place a thermometer to measure the entire earth's temperature. During the last few centuries, local weather agencies established what has become a network of temperature-collection stations scattered throughout the world, such as the one pictured in Figure 2-1. Groups collecting these data created standards, such as how high above the ground measurements were taken, and specified the use of ventilated enclosures. Ships and buoys made measurements of air temperatures above the ocean. During the 1940s, explorers ventured into the polar regions to collect temperature readings that are now represented routinely in the global temperature measurements.

New measurement locations were added continuously to the database, and measurement techniques became more refined. As this happened, scientists did not want the evolution of the measurement network itself to affect the consistency of the global averages they were trying to measure. To prevent this, the groups tracking the data incorporated the use of *anomalies*, which refer to departures from an established local temperature range. The change in temperature compared with some previous level is called an *anomaly*. Uncertainties in measurement are most likely to affect the absolute temperature reading—or the actual number of degrees a particular station was reading. Tracking anomalies makes it easier to answer the overriding question of how the temperature is changing at a given location. Scientists then can talk about how much a particular reading is above or below normal. Anomaly tracking is also used to reduce the possibility of calibration errors with ocean temperature monitoring and with sea-level measurements, as we will see in the following chapters.

Today, over 7000 stations spanning the earth, over both land and the oceans, perform temperature measurements (Figure 2-2).

Three major organizations collect and compile temperature data to generate global averages:

Figure 2-1 An early temperature-measurement station in Utah from around 1930. (*Source: NOAA.*)

1. The United States National Oceanic and Atmospheric Administration (NOAA)
2. The United States Aeronautics and Space Administration (NASA)
3. The Climate Research Unit (CRU) at the University of East Anglia in the United Kingdom

If a major league batter goes down swinging twice in a row, we might call it a slump and forget about how many runs he batted in the month before. Before the team's manager decides to send that player back to the minors, he would be well advised to check the player's batting average year to year. Similarly, we may be tempted to jump to a conclusion about the earth's climate based solely on a record hot day, an unusually warm winter, a devastating series of storms, or a drought in one part of the world and flooding in another. However, climate typically changes slowly in ways that are hard to notice unless we look at it from the vantage point of long-term global averages.

Figure 2-2 Temperature-measurement station. (*Source: NOAA.*)

How Is the Earth's Temperature Changing?

RECENT HISTORY—THE PAST 150 YEARS

Direct thermometer-based readings from around the earth for the past 150 years show that the earth is getting warmer. This is reflected in the graph shown in Figure 2-3. This period is called the *instrumental record* because all the data are

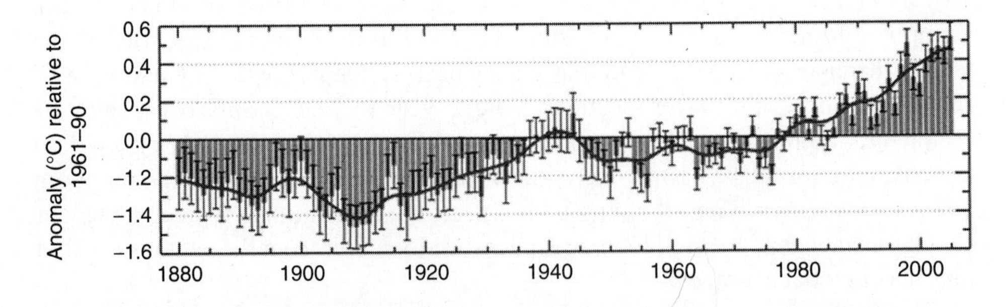

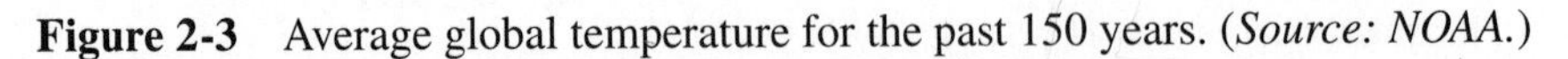

Figure 2-3 Average global temperature for the past 150 years. (*Source: NOAA.*)

derived from direct thermometer measurements from locations around the world.

A hundred years ago, the average temperature of the earth was about 13.7°C (56.5°F); today, it is closer to14.4°C (57.9°F). At first, this may not seem like a very large change. But the earth's temperature usually takes many centuries to change by as much as a degree. With the earth, small changes—or at least small changes from our perspective of day-to-day weather extremes—can have significant consequences. The last ice age was, on average, only about 5°C (9°F) colder than present-day global averages. A past interglacial warm period, during which sea levels were about 4–6 m (13–20 ft) higher than present levels, was, on average, about 5°C (9°F) warmer than today. At the rate that the earth is warming currently, substantial global climate changes very well may have been set in motion.

During the past 150 years, the earth has slowly become warmer (with some ups and downs), mostly between the years 1910 and 1940. Not much change occurred prior to that, from 1850 to 1910, except for minor ripples primarily from small natural variations and possible inconsistent sampling. From 1940 to 1975, there was a slight cooling trend, probably related to increased sunlight reflecting from the atmosphere, as industrialization evolved along with the air pollution that it generated following World War II.

Beginning with the 1970s, the pace picked up. The average global temperature increased more rapidly—at a rate of 0.2°C (.36°F) per decade. The warmest years on record are the most recent.

Because of the global nature of this problem, thousands of scientists from over 30 countries have formed an organization called the *Intergovernmental Panel on Climate Change* (IPCC). Members of the IPCC, coordinated through the United Nations, have been collaborating since 1988 to interpret data relating to climate change. In 2007 The IPCC, along with Al Gore, was awarded the Nobel Peace Prize for their efforts in studying climate change. Their most recent findings were released in a report called "Climate Change 2007, the IPCC's Fourth Assessment Report" (AR4).

Some official results of that report are

- The average global temperature has increased by 0.74°C (1.3°F) over the past century.
- Eleven of the warmest years on record have occurred during the most recent 12 years of the study (1995–2006). This indicates that not only is the earth's average temperature rising but also that the rate of global warming is also increasing.
- The warming trend for the last 25 years is more than double that of the past century (Figure 2-4).

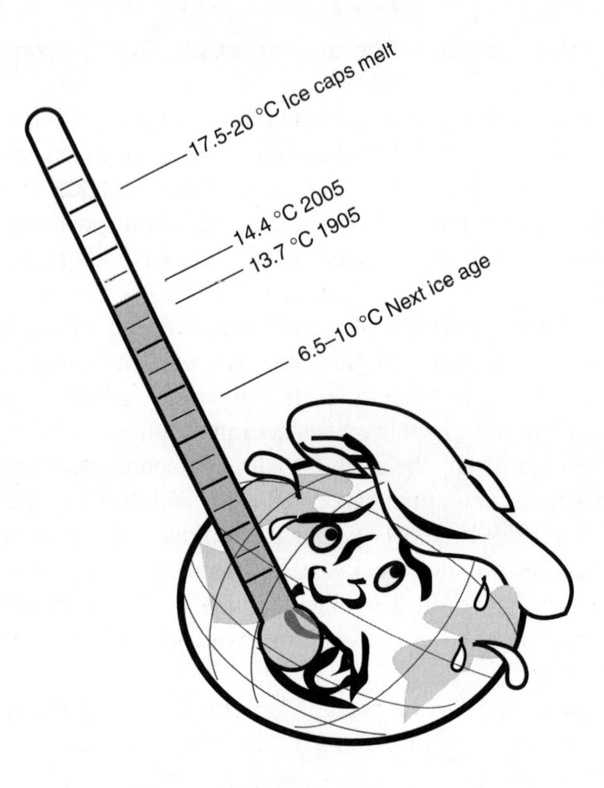

Figure 2-4 Earth with a fever.

> How much should we be concerned with a temperature rise of less than 1°C? The last ice age was, on average, only about 5°C (9°F) colder than present-day global averages.
> The last time the global average temperature was about 5°C (9°F) warmer than today, sea levels were 4–6 m (13–20 ft) higher than present levels.

THE PAST 13 CENTURIES

Since the cave men did not have thermometers thousands of years ago, scientists today have to rely on indirect methods to determine what the earth's climate once was. Scientists need to work more like crime scene detectives to discover what the earth was like in the past. Today, scientists carefully search for obscure pieces of evidence left buried in the snows of Alaska, among centuries-old debris on the

ocean floor, or within multiple layers of coral. As part of this forensic effort, scientists carefully compare the different types of atoms that may have been present in a geologic sample thousands of years old.

What they are finding is that the most recent 50-year period in the northern hemisphere is warmer than it has been for the previous 1000 years (Figure 2-5). What is happening currently is statistically unusual and cannot be considered simply part of the normal ebb and flow of natural cycles. The earth's temperature is off the charts.

This chart takes the form of a "hockey stick," which in statistics is typically associated with an abrupt change or a breakout pattern. The last century is that part of the hockey stick that contacts the puck, and the previous 1200 years represent the stick.

The darker line on the right hand side of the graph is the instrumental record, which includes calibrated thermometers and, more recently, satellite measurements. Reliable direct temperature measurements can only take us back a little more than a century and a half. These are the same data that are reflected in Figure 2-3. The instrumental record inherently has a greater degree of precision than the indirect (paleontologic) records.

The multiple lines on the left show data derived from a variety of techniques used in this thermal detective work, including tree-ring patterns, ice-core sample composition, and coral reef growth-band analysis. (We will discuss how this is done a little later in this chapter).

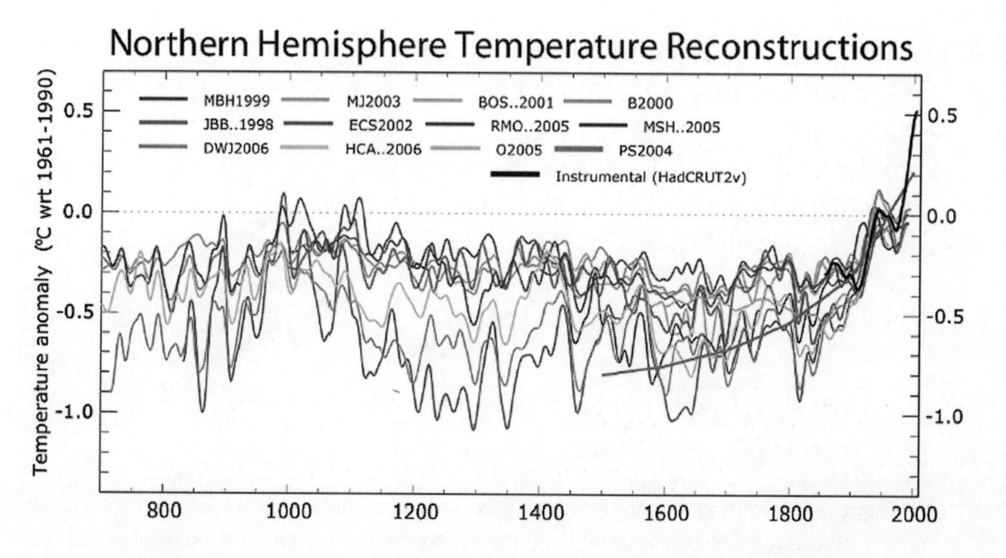

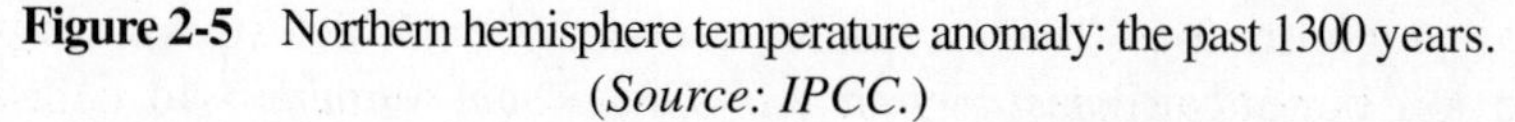

Figure 2-5 Northern hemisphere temperature anomaly: the past 1300 years. (*Source: IPCC.*)

To compare apples with apples more easily, the temperature anomalies rather than the actual temperatures are shown. A baseline is established for all the data for the time period 1961–1990 (where the anomaly is defined as zero). This graph then shows how much greater or lesser the temperatures for a particular year are compared with the baseline period.

ANCIENT HISTORY

Ice-core, tree-ring and other paleontologic studies suggest that the warm temperatures that are being measured currently are unusual and have no precedent during the previous 1300 years. The last time that the polar regions were warmer than the present temperature levels was more than 125,000 years ago. It is also noteworthy that, at that time, reductions in polar ice volumes resulted in a significant rise in the sea level and relocation of many coastlines from their present-day positions.

The earth went through a number of temperature excursions during its formative years over the past several hundred thousand years. Scientists (called *paleoclimatologists*) study this indirect historical climate record that Mother Nature has conveniently left behind for us to discover. Like crime scene investigators, scientists reconstruct historical climate conditions by examining evidence found in tree rings, by studying the layers of ocean sediment, and by analyzing the composition of layers taken from ice cores. Figure 2-6 shows a pattern of temperature changes that corresponds to four climate cycles going back 450,000 years.

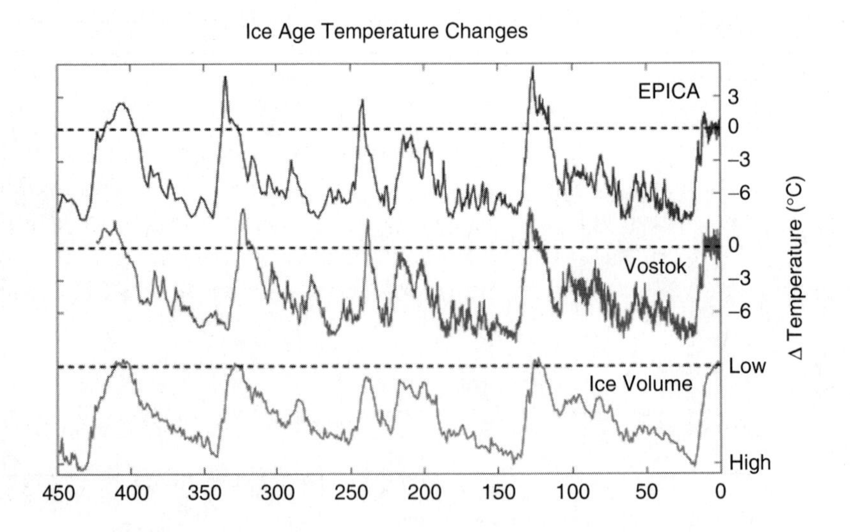

Figure 2-6 The last four ice ages (the past 450,000 years). *(Used by permission from R. Rohde, Global Warming Art.)*

The measurements were made at two separate sites in Antarctica—one at Vostok Station and the other at a separate location by the European Project for Ice Coring in Antarctica (EPICA). Both sites show that earth's temperature has never deviated from a temperature range of roughly 10°C (18°F). We are currently about a degree or two above the midpoint of that range. This puts into perspective the nearly 1°C change that has already been determined for the past century with direct temperature measurements and the rate of increase that has been suggested by the trend in recent years. If the average global temperature continues to rise at the same rate as has occurred during the past decade, we may see climatic changes similar to those that have accompanied the retreat of past ice ages. Four additional ice age cycles moving through a similar temperature range and going back 650,000 years were obtained from ice-core data.

WHAT IS "NORMAL" FOR THE EARTH?

Temperatures have gone through natural cycles in the past. The medieval times were warm, followed by a cooler period from the seventeenth through the nineteenth centuries, after which a warming trend occurred. However, at no time during the past 11,600 years (the Holocene era in geologic time) have temperatures been as high as they are today.

HAS GLOBAL WARMING EVER HAPPENED BEFORE?

The answer to this question is yes. However, before we take comfort in that thought, we should note that sea levels were most likely 4–6 m (13–20 feet) higher then than they are today.

Is this recent warming trend simply a phase that the earth is going through? Or is this something different? Looking back through the geologic records, we find periods of cold in the form of ice ages. These periods of cold alternated with interglacial warm periods. In some cases, the temperature records show that there were periods in history where the earth's temperature was greater than it is today and greater than it would be if it continues to heat up at current rates (Table 2-1).

Going Back in Time—The Next Best Thing to Having a Thermometer

Let's take a closer look at how scientists gather information about Earth's thermal history. The indirect methods they use are called *proxy measurements*.

Table 2-1 A Brief History of Climate Change

Period	Time Period	Characteristics
Instrumental record (direct measurements)	Past 150 years	On the warm side of the historical range and increasing
Little ice age	600–150 years ago	Coldest in thousands of years, possibly caused by volcanic activity and reduced solar activity
Medieval warming	1050–750 years ago	Warm—close to modern-day levels—drought in North America
Interglacial warm period	11,000 years ago	Warm—ice sheets retreated
Younger Dryas period	13,000–11,700 years ago	Cold followed by abrupt warming
Ice ages	Four events between about 1 million and 115,000 years ago	Ice sheets covered much of Canada and northern Scandinavia—advanced and retreated four times
Interglacial warm period	3 million years ago	Much warmer climate with reduced global ice cover and higher sea levels

ICE-CORE SAMPLES

Scientists collect samples from buried ice layers (Figures 2-7 and 2-8) that have been left undisturbed in polar regions for thousands of years. Ice-core drilling began in the 1960s in the ice sheets of Greenland and Antarctica. A core sample over 3053 m long (nearly 2 miles) was removed recently from the Greenland ice sheet, providing a climate record of over 100,000 years. This was known as the Greenland Ice Sheet Project 2 (GISP2). It took five summers to extract the core sample. Pieces of this
ed at –20°C for a year following extraction to allow for
fter being exposed to the intense pressure under the ice
he layers in these ice-core samples were painstakingly
the National Ice Core Laboratory in Denver, Colorado,
of temperature, precipitation, atmospheric composition,
patterns. Visual records from Antarctic ice go back even
000 years.
determine the climate conditions that existed at the time that those layers formed. As with the rings of trees (which provide similar information but over a much shorter span of time), each year produces an identifiable layer, as can be seen in Figure 2-9. Snow falls in the polar regions, adding to the layers year after year. Summer snow often has a different crystalline structure than winter snow. In winter, the solid layer sublimes and then redeposits, forming a layer that can be distinguished by the texture of the ice. Layers of particulates called *aerosols* that fall out of the air provide an additional marker for each annual layer. Ice cores

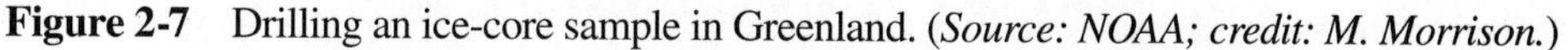

Figure 2-7 Drilling an ice-core sample in Greenland. (*Source: NOAA; credit: M. Morrison.*)

provide a vertical timeline of past climates in glaciers and ice sheets that are preserved for as long as hundreds of thousands of years.

Scientists have developed a method to measure small amounts of trace elements that occur naturally in the earth's atmosphere. By means of this technique, they determine historical temperatures using various chemical elements called *isotopes*. One such isotope is a form of the element oxygen called *oxygen-18*. This isotope is chemically identical to the much more common oxygen-16. However, oxygen-18 (having two extra neutrons in its nucleus) is slightly heavier. During the years when the earth is warmer, more of this form of oxygen gets incorporated into the ice (which is water, or H_2O). By comparing the amount of heavy oxygen with the amount of normal oxygen, it is possible to unravel the changes in temperature that have occurred over time. A good example of this is given in Figure 2-8. The pattern is one of alternating warm and cool periods resulting in glacial and interglacial cycles. More information on isotopes can be found in Chapter 5.

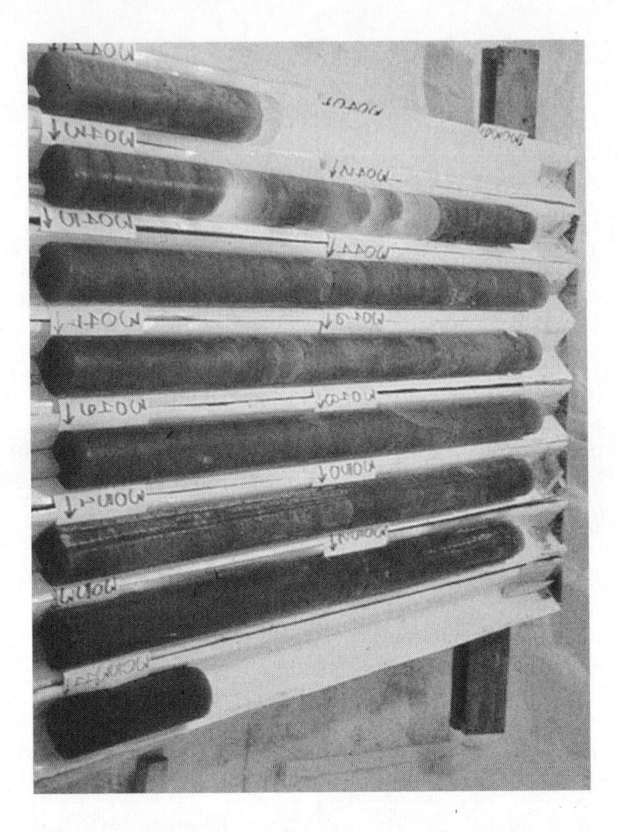

Figure 2-8 Sections of GISP2 ice core being prepared for analysis. (*Source: NOAA.*)

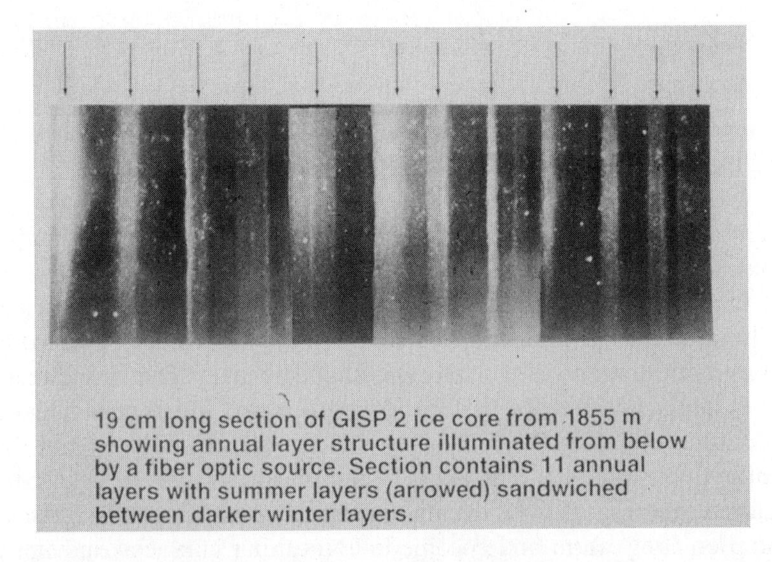

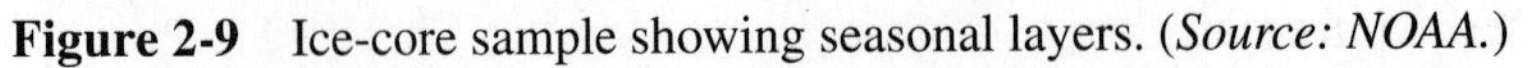

Figure 2-9 Ice-core sample showing seasonal layers. (*Source: NOAA.*)

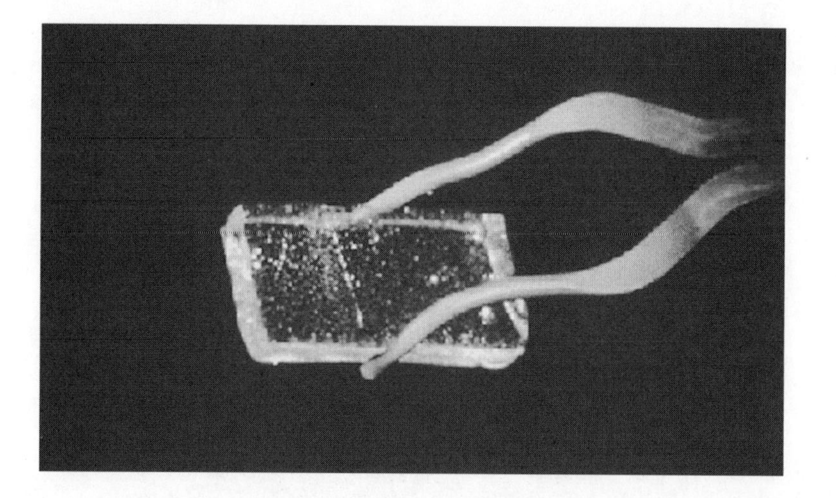

Figure 2-10 Atmospheric gas from the past can be trapped in ice-core samples. (*Source: NOAA; credit: M. Morrison.*)

Paleoclimatologists take this investigation even further. Small bubbles of atmospheric gases get incorporated into the layers that form the ice-core samples (Figure 2-10). Scientists analyze the composition of these gases and correlate their occurrence with the temperature changes. Rises and declines of the critical greenhouse gases—carbon dioxide, methane, and nitrogen dioxide—have been found to correlate with the major global temperature changes in the past. As mentioned in Chapter 7, carbon dioxide is thought to have accelerated warming trends set in motion by other causes rather than act as the primary cause of past global warming.

TREE-RING SAMPLES

It is common for students determine how old a tree is by counting the rings of a cross section cut from the tree's trunk. Each year the tree adds a new ring that reveals clues about the climate at the time the tree was growing. A wider ring would suggest a warmer growing season. Fortunately, researchers can withdraw a core the diameter of a pencil from a tree without killing it (so that the tree can, among other things, continue to remove carbon dioxide from the air). Tree-ring data do not go as far back in time as some of the other proxy temperature measurements. However, tree-ring data provide excellent detail in distinguishing individual yearly events. This can be seen in Figure 2-11. Analysis of tree rings helps to put a more accurate time stamp on data derived from some of the other techniques for time periods where they overlap.

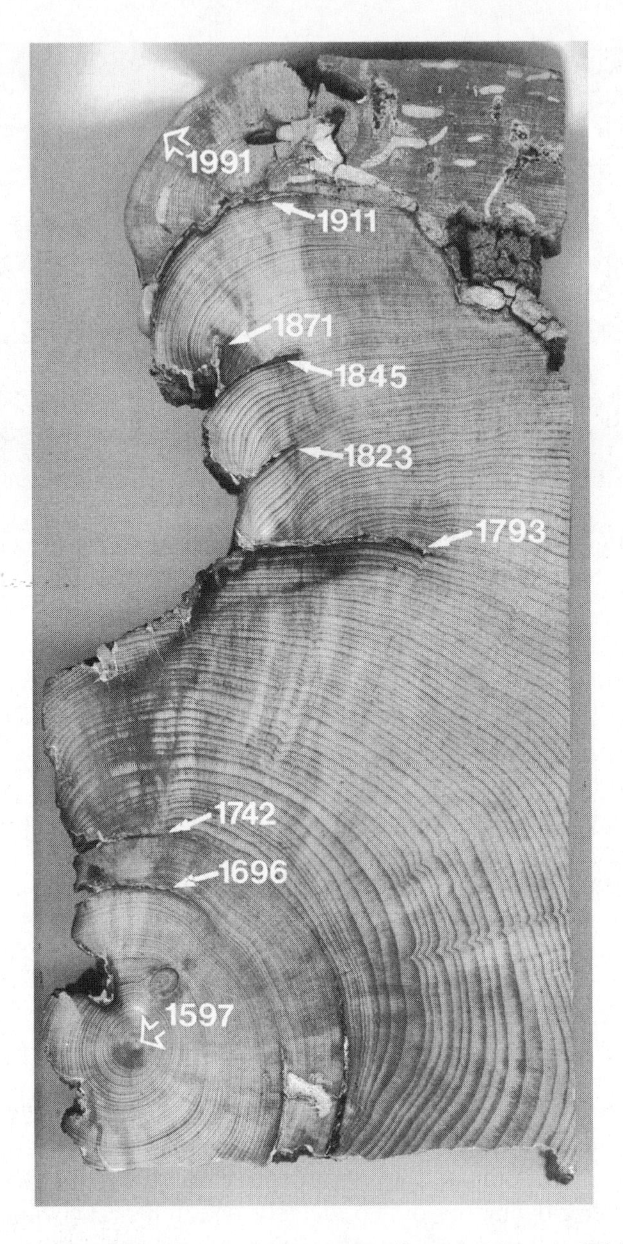

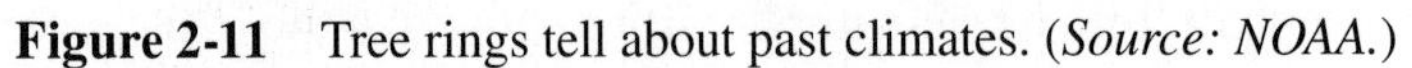
Figure 2-11 Tree rings tell about past climates. (*Source: NOAA.*)

CORAL GROWTH

Coral provides a unique perspective on past conditions in the ocean and offers a look at what may have occurred well before the instrumental record. Core samples are extracted for analysis, as shown in Figure 2-12.

Coral is a tiny sea animal whose skeletal remains are formed from calcium carbonate (Note that each coral is an individual animal and the reef is composed of their skeletons.) During the winter, skeletons that form in coral reefs have a different density than those formed in the summer because of variations in growth rates related to temperature and other conditions. Core samples taken from coral reefs exhibit seasonal growth bands that can be analyzed in a similar manner as those observed in trees. The different layers and their association with temperature changes are shown in Figure 2-13.

Figure 2-12 Extracting a core with a hydraulic drill on a coral colony at Clipperton Atoll. (*Source: NOAA.*)

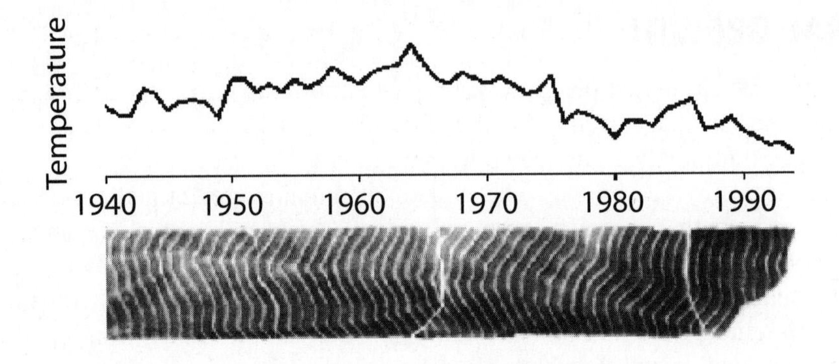

Figure 2-13 Coral section showing temperature variations with time. (*Source: NASA; credit: R. Simmon.*)

BOREHOLES AND OCEAN SEDIMENT

In most places around, the world ice layers are not available to provide historical climate data. To provide a more complete picture geographically, scientists study rock samples drilled from the earth's crust. This type of borehole data is only able to detect changes from one century to another rather than the higher time resolution that comes from other proxy techniques. Borehole measurements confirm that the twentieth century was the warmest of the past five centuries and provide a basis for comparison from one region to another.

Each year, dust, plants, and animal skeletons settle on the ocean floor. Layers of sediment form a vertical record of past climates. In a similar manner, rock sediments reveal a timeline in continental areas, and careful extraction of vertical samples permits scientists to go back in time by analyzing the sediment layers.

Different Places, Different Temperature Changes

NORTHERN AND SOUTHERN HEMISPHERES

Land regions have shown greater warming than the oceans in the past few decades. This observation is consistent with the greater heat capacity of water compared with air. Since the northern hemisphere has a greater amount of land compared with water, it has shown a more definitive rise in temperature compared with the southern hemisphere. Historical records for the southern hemisphere are too limited to provide a similar comparison of recent temperatures and past climate patterns.

Global warming has been highest at the higher northern latitudes, with Arctic temperatures increasing at almost twice the rate of the rest of the earth during the past century. This may be the result in part of long-term changes in atmospheric and ocean currents that redistribute absorbed heat.

Averages over time or across the entire planet provide a useful way to simplify an otherwise apparently chaotic pattern. However, sometimes important information can be lost in the statistics. For instance, when the average temperature of the entire planet increases by 1°C some places may be warmer that year and other places colder. Usually, the north pole heats up nearly twice as fast as the equator. Although there has been a warming trend, there have been recent years when for various reasons the earth was cooler. The warming trend has not affected the Antarctic region nearly as much as the northern latitudes.

URBANIZATION: THE HEAT ISLAND EFFECT

The process of urbanization has led to concentration of population in cities. Structures in urban areas tend to trap heat, resulting in higher local temperatures on a given day in a process called the *heat island effect*. This causes urban air to be about 1–6°C (2–10°F) hotter than that in surrounding rural regions. However, according to the IPCC, heat island effects have a negligible influence on average global temperatures (less than 0.006°C per decade over land and zero over the oceans). The heat island effect may make people feel more uncomfortable on sultry summer evenings, but it is not considered to be a factor in accelerating the melting of the polar ice caps.

The heat island effect is caused by

- Reduction of nighttime radiation of heat absorbed during the day. The concrete and steel structures that constitute the city act as a blanket that keeps the heat in that otherwise would escape overnight.
- Concentration of light that reflects from vertical surfaces.
- Restriction of conductive and convective pathways for heat to escape in what are effectively urban canyons.
- Local concentration of greenhouse gases result in greater absorption of heat from the earth.
- Local heat generation from cars, air conditioners, and industry.

More than 50 percent of the world's population live in urban areas, with this number approaching 75 percent in Western countries. As a result, many people are experiencing an increase in local temperatures that may affect their health and comfort but that is unrelated to global warming. To avoid corrupting global temperature measurements, temperature stations that are considered to be urban are excluded from the data set, and efforts are made to ensure that what is being looked at as a global phenomenon is truly global in nature.

VERTICAL TEMPERATURE CHANGES

Radiosonde Balloons

Since the late 1930s, the National Weather Service (NWS) has been making air observations using *radiosondes*. These are small, expendable instrument packages that are carried into the atmosphere by a balloon filled with hydrogen or helium (Figure 2-14). The radiosonde measures profiles of pressure, temperature, and relative humidity. Scientists gather information on wind speed and direction that is useful for weather forecasting and predicting severe weather events. When the balloon has expanded beyond its elastic limit, which usually occurs at about 30,000 m (almost 100,000 ft), it bursts. A small parachute slows the descent of the radiosonde to enable it to be reused.

Figure 2-14 Radiosonde balloon. (*Source: NOAA.*)

There are over 800 upper air observation stations in use throughout the world, the majority being in the northern hemisphere. Through international agreements, data are exchanged between countries. Although the primary use of radiosonde balloons is for shorter-term weather purposes, the data also are being correlated with ground and satellite temperature measurements.

All three temperature-measurement methods (i.e., ground temperature, radiosonde, and satellite) agree that both the surface and the troposphere have warmed and that the stratosphere has cooled. Table 2-2 lists the layers of the atmosphere. Early radiosonde data, however, were not consistent as to whether the *rate* of warming of the troposphere is greater or less than the surface warming. Errors in radiosonde measurements have been identified and corrected. Current radiosonde data are now consistent with satellite and ground-based measurements.

Satellites

Low Earth orbit satellites have been measuring the earth's temperature from space since 1979. Satellites in polar orbit (such as the one pictured in Figure 2-15) circle the earth approximately every 90 minutes.

With each pass around the earth, the orbit of these satellites follow a slightly different path, providing nearly complete coverage of the entire surface of the earth. By recording the components of the (invisibly) glowing earth, scientists can measure the temperature of the Earth from space. The instrument that is used on most of the weather satellites is called the *microwave sounding unit* (MSU). The MSU records heat radiated from the earth in the form of invisible infrared waves, which are then interpreted as temperature.

Table 2-2 Layers of the Earth's Atmosphere

Layer	Height Above the Earth's Surface	Approximate Temperature Range	Key Characteristics
Thermosphere	To about 600 km (372 miles)	Over 1700°C	Very low density of gases; sometimes called the *ionosphere*
Mesosphere	To about 85 km (53 miles)	To –93°C	Temperature decreases until a boundary called the *mesopause*
Stratosphere	To about 50 km (30 miles)	To –3°C	Contains ozone layer; warmer because of ultraviolet light absorption
Troposphere	8–14.4 km (5–9 miles)	17 to –52°C	Layer where all weather occurs; troposphere's height varies with solar energy; it is lowest at the poles and highest at the equator

Figure 2-15 Satellites in orbit measure the earth's temperature using instruments that collect long-wavelength light energy. (*Source: NOAA.*)

One of the main advantages of satellites is that they monitor almost the entire earth each day, providing the largest statistical database and the greatest resolution of variation from one place on the earth to another. Satellites measure the total heat radiated from the earth, which has a direct bearing on the earth's temperature. Satellites, however, are limited in their ability to separate temperatures at different atmospheric layers, although some altitude information can be extracted from the data. Satellites collectively carry 13 different instruments. These also provide a profile of temperature from the earth's surface up through the troposphere (from the surface to about 10 km), which is the lower layer where all the earth's weather occurs and the stratosphere above it (from about 10–30 km).

Scientists working with MSU data encountered some initial problems in interpreting the precise details of the vertical temperature profile through the layers of the atmosphere. The IPCC's Third Assessment Report in 2001 identified discrepancies with the early satellite data related to refinements needed to establish precise orbital positions. These discrepancies have been corrected and the satellites data now agree with measurements derived from radiosonde balloons and near-surface thermometers:

- The average global temperature is increasing and is warming at an increasing rate.
- The troposphere is warming at a rate that is slightly more rapid than at the surface. This is consistent with models that attribute the atmospheric

warming to absorption by greenhouse gases. The troposphere actually appears to be getting larger as a result of atmospheric warming.

- The stratosphere is cooling. With more heat extracted from the atmosphere at lower altitudes, there is less heat left at higher altitudes. Depletion of ozone in the stratosphere also contributes to lower temperatures there.
- Land areas are heating at a more rapid rate than the ocean. This can be at least partially understood based on the fact that the heat capacity of water is more than four times that of air. This means that for a given amount of heat absorbed, water will increase only one-quarter the number of degrees of an equal mass of air.
- The northern hemisphere is heating at a much more rapid rate than the southern hemisphere. The temperature increase is Alaska is nearly four times as great as the global average.

What the Data Are Telling Us

When the IPCC released its previous report (AR3) in 2001, much of the discussion at the time in the media and various publications referred to "the debate on global warming." Although there remain open issues and ongoing investigations, a broader consensus is now emerging.

The IPCC presented the following conclusions in its 2007 report (AR4):

- Warming of the climate system is "unequivocal."
- There is a "very high confidence" that human activities since 1750 have contributed to this net warming of the earth.
- The increase in the concentration of carbon dioxide in the atmosphere is the single most important factor driving the increase in global temperatures.
- If tomorrow, carbon dioxide emissions into the atmosphere were to stop completely, carbon dioxide levels already in the atmosphere will cause an average sea level rise of about 1.4 meters (4.6 feet).
- By as early as 2030, 75 million to 250 million people in Africa will suffer water shortages.
- Residents of many of Asia's large urban areas will be at great risk from river and coastal flooding.
- Extensive loss of the species is expected especially in Europe.
- In North America, longer and hotter heat waves and greater competition for water resources are forecast.

What Is the IPCC?

Much of the climate modeling is developed, reviewed, and reported by the IPCC.

- The IPCC was established by member countries of the United Nations in 1988.
- Its function is to assess scientific data relating to climate change. Over 2,500 scientists from around the world contribute to the IPCC's assessment efforts.
- The IPCC published overall assessments in 1990, 1995, 2001, and 2007.
- Three separate working groups were set up to focus on the physical science of climate change, the impact on people and nature, and methods to correct the problem.
- The results of the study groups are peer reviewed by hundreds of reviewers from over 130 countries.
- In 2007 the IPCC along with Al Gore won the Nobel Peace Prize for their work on climate change.

Cause and Effect

Temperature measurements from around the world indicate that the temperature of the earth is increasing in a way that is unprecedented in recent times. What is causing this to happen? How much is natural, and how much is the result of human activity? Will it be a big deal? Will it be catastrophic? Is there anything we can do to reverse the course that has apparently been set in motion? These questions will be addressed in the sections of this book that follow.

Key Ideas

- Reliable direct temperature measurements of the earth's air temperature have been made for roughly the past 150 years. This is known as the *instrumental record.*
- Direct measurements show that the average global temperature has increased by 0.74°C (1.3°F) during the past 100 years. At present, the temperature is increasing at a rate of 0.177°C (0.211°F) every 10 years.

- The average global temperature of the earth is now 14.4°C (57.9°F). One hundred years ago, the average global temperature was 13.7°C (56.5°F).
- Eleven of the past 12 years were the hottest on record.
- To compare present-day temperatures with temperatures during historical periods, scientists use indirect (or proxy) methods to determine temperatures in the past.
- Proxy methods include ice cores, tree rings, coral layers, ocean sediment, and boreholes.
- The northern hemisphere is warming faster than the southern hemisphere.
- The heat island effect causes locally higher temperatures but is not mistaken for and is not a cause of global warming.
- Satellites have determined that the troposphere is warming slightly more rapidly than at the surface, which is consistent with the concept of greenhouse gas absorption of heat.

Review Questions

1. What is meant by a temperature anomaly?
   (a) Temperature extremes
   (b) A sudden change in a temperature
   (c) The amount that a temperature reading varies from an established baseline
   (d) An error in a temperature reading
2. Approximately how much did the average global temperature increase over the past 100 years?
   (a) 0.25°C (0.5°F)
   (b) 0.75°C (1.4°F)
   (c) 2.0°C (3.6°F)
   (d) 5.0°C (9.0°F)
3. Which of these is a proxy method of characterizing temperature?
   (a) Coral-core layer analysis
   (b) Weather station at Big Sky Airport in Phoenix
   (c) Air measurements outside of Anchorage, Alaska
   (d) Rough estimates of air temperature using an uncalibrated thermometer

4. How does the greater atomic mass of oxygen-18 make it useful in determining prehistoric temperatures in ice-core samples?
   (a) Oxygen-16 changes into oxygen-18 at higher temperatures.
   (b) Oxygen-18 is released from water molecules at higher temperatures.
   (c) Oxygen-18 is slightly heavier and is incorporated more into an ice layer during warm weather.
   (d) Water with oxygen-18 freezes at a lower temperature than water with oxygen-16.
5. What is the primary function of the IPCC?
   (a) To collect temperature data
   (b) To perform measurements on ice cores
   (c) To perform research on ways to reduce greenhouse gas emissions
   (d) To analyze data
6. How does the temperature change over land areas compared with the change of ocean temperature?
   (a) Lower
   (b) Higher
   (c) Same
   (d) Mixed bag
7. According to GISP2 ice-core data, about how much warmer were previous interglacial warm periods (when ice caps had melted) than today?
   (a) 0.75°C (1.4°F)
   (b) 1.5°C (3.6°F)
   (c) 5.0°C (9.0°F)
   (d) 10.0°C (2.7°F)
8. Where is the atmospheric temperature increasing most rapidly?
   (a) Alaska
   (b) The equator
   (c) The Pacific Ocean
   (d) Northern Europe
9. How are satellites able to measure the temperature of the atmosphere?
   (a) By measuring how much light is reflected from clouds
   (b) By measuring invisible electromagnetic energy radiated from the earth

(c) By sending a microwave signal and measuring how much is reflected back

(d) By observing the movement of clouds and ocean currents

10. How are temperatures changing vertically in the atmosphere?

(a) Both the troposphere and the stratosphere are warming.

(b) Both the troposphere and the stratosphere are cooling.

(c) The troposphere is cooling and the stratosphere is warming.

(d) The troposphere is warming and the stratosphere is cooling.

CHAPTER 3

Signs of Global Warming

Direct and Indirect Evidence

All indications are that the earth's atmosphere is warming. We might expect to observe other signs around the world that are consistent with this conclusion. Global ocean temperature and sea level are beginning to rise in small but measurable amounts. Rapid melting of snow and ice observed in many parts of the world provides a more dramatic signal that global warming is underway. This chapter will explore the signs of warming that are beginning to show up in various places around the world.

Here are a few brief examples:

- *Glaciers and ice caps.* Each year, enough fresh meltwater is brought into the world's oceans to raise sea level by 0.8 mm.
- *Snow cover.* Every 10 years, there is 2 percent less snow cover in the northern hemisphere than there was a decade before.

- *Lakes and rivers.* Each year, the world's lakes and rivers are covered by ice 12 fewer days during the winter months than they were150 years ago.
- *Permafrost.* There is 7 percent less permanently frozen areas (*permafrost*) than there was in 1900.
- *Spring.* Leaves unfold, birds migrate, and birds lay eggs earlier every year.

The Oceans

OCEAN TEMPERATURE

Sea Surface Temperature

As the temperature of earth's atmosphere goes up, the average *sea surface temperature* is also rising. The *sea surface* refers to the top few meters of the ocean. Data are derived from satellites and through a global network of buoys such as the one pictured in Figure 3-1. Global average sea surface temperature differences (anomalies) compared with a base period defined as from 1961–1990 are shown in Figure 3-2.

The warmest year since these measurements began in 1856 was 1998. The five warmest years have occurred since 1995. Currently, the mean global sea surface

Figure 3-1 Buoy deployed at sea to collect ocean temperature data. (*Source: NOAA.*)

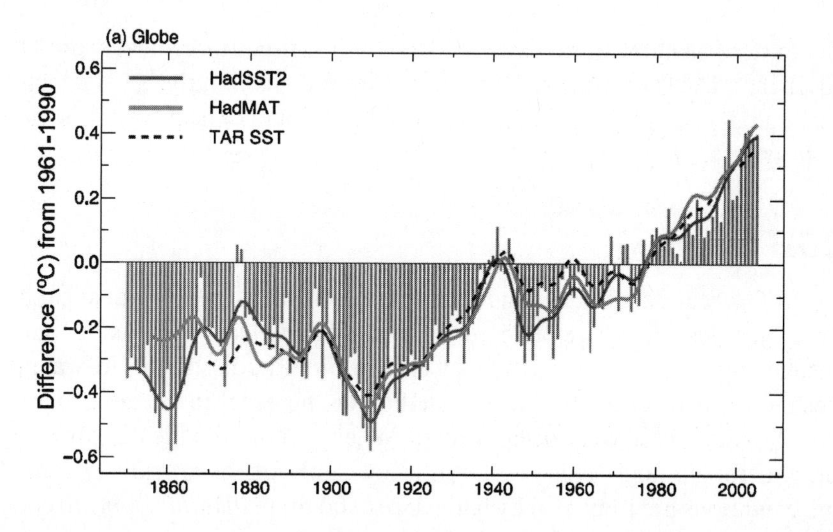

Figure 3-2 Global average sea surface temperature anomalies from 1850 to the present. (*Source: IPCC.*)

temperature is about 0.35°C (0.6°F) above the 1961–1990 baseline. The sea surface is showing a similar pattern of increasing temperature as the atmosphere. Notice that in the early 1940s there were several warmer years. This is most likely the result of a series of El Nino events.

Most regions of the ocean have warmed with a few localized exceptions, such as the southern coast of Greenland. It is also important to note that the small changes in ocean temperature are consistent with the quantity of additional heat that is being absorbed by the earth.

Where the Heat Goes

Since 1960, the oceans have absorbed 20 times more heat energy than the atmosphere. As we will see in Chapter 4, nearly 90 percent of the sun's energy goes into the earth's oceans. However, if we consider a layer of the ocean to a depth of 700 m (2300 ft), we find that the average temperature increase in that layer is much smaller. The average global ocean temperature in an ocean layer 700 m deep has increased (from the period 1961–2003) by a meager 0.1°C (0.2°F). The oceans absorb an enormous amount of heat. But heat is not the same as temperature. It takes much more heat to change the ocean temperature than it takes to change the air temperature. The ocean gets more heat from the sun but shows a more modest temperature change in response because of its greater heat capacity.

Had the oceans not been able to store as much heat as they do, the temperature of the atmosphere would have been much greater. Because of their greater size and

ability to store heat, the oceans can hold a thousand times more heat overall than the atmosphere. The oceans are enormous, so the seemingly small increase in temperature represents a massive increase in the amount of heat energy in the earth's climate system.

ACIDITY

The world's oceans are becoming more acidic each year. The more acidic the oceans become, the less they are able to continue taking carbon dioxide out of the atmosphere. This is a direct result of carbon dioxide dissolving in water, which makes the water more acidic. As carbon dioxide increases in the atmosphere, more carbon dioxide dissolves in ocean waters. When carbon dioxide dissolves in water, it forms carbonic acid. This results in a lower pH of the world's oceans. (More *acidic* conditions are indicated by a lower *pH* and more *alkaline* or *basic* conditions are indicated by a higher pH level.)

Biologists do not yet fully know the effect of increased pH on aquatic organisms. A typical pH range for seawater around the world is 7.9–8.3, which is slightly alkaline, or basic. Since 1750, the average pH range of the global oceans has decreased by 0.1, indicating a slight shift to being more acidic.

There is no evidence that the pH was ever more than 0.6 below the preindustrial levels during the past 300 million years. Scientists are concerned about potential impacts on sea life from decreased ocean pH level, especially if it continues to fall. The Intergovernmental Panel on Climate Change (IPCC) predicts that increasing carbon dioxide levels in the atmosphere will raise ocean pH by 0.14–0.35 during the twenty-first century. This is in addition to the 0.1 increase seen already.

To determine how ocean pH may affect the ocean ecosystem, scientists are using satellites to monitor the health of the oceans. Satellites are beginning to measure the extent of basic aquatic plant activity (in the form of phytoplankton consuming carbon dioxide, releasing oxygen, and producing food) in the ocean. *Primary production* is the basic ecologic process in which carbohydrates are produced by plants through photosynthesis. This forms the basis of the ocean food chain. Satellites measure primary production by monitoring the very specific green wavelengths of visible light that indicate the presence of chlorophyll. Initial indications are that there has been a 6 percent decrease from the 1980s to the 1990s. These data carry too much uncertainty to be a cause for alarm, but they are something that scientists will continue to monitor.

SALINITY

When large masses of ice melt, they add freshwater to the oceans. One way to tell if large masses of ice are melting is to monitor the salinity of adjoining areas of the seas. The oceans are about 3.5 percent salt—mostly sodium chloride, with some other "salts" such as magnesium sulfate mixed in. The salt content of the oceans is a good chemical signature that can identify flow patterns such as ocean currents. The boundaries of the Gulf Stream can be distinguished by abrupt changes in salinity. A lower salt concentration can mean that freshwater such as from glacial melt or increased precipitation is coming into seawater. A higher salt concentration can mean that above-average evaporation is removing water and leaving a more salty mix behind.

Based on millions of measurements, scientists are finding that the oceans are freshening. This is a growing concern because the great ocean currents such as the Gulf Stream depend on salinity differences to keep them moving. Modifications to the salinity patterns in the ocean may affect these currents in ways that are hard to predict. Some scientists believe that abrupt climate changes may have been triggered by the sudden release of fresh meltwater in the past and that this may happen again.

SEA LEVEL

How Sea Level Is Measured

Sea level around the world has increased by a small but measurable and statistically significant amount. It is not enough of an increase to affect any coastal areas or to even be noticeable. It is premature to look for consequences of sea level rise just yet. But scientists are now able to detect small changes in the sea level—both from sea gauges and from satellites in earth orbit.

With the beginning of the nineteenth century, the global average sea level, according to gauge measurements, began to rise at a slow but steady rate of 1.7 mm (only slightly more than $^{1}/_{16}$ in) per year. Archaeological records from the Mediterranean region suggest that sea level had been fairly stable for the 2000–3000 years prior to about 1870. The seas already had risen by about 120 mm (nearly 4 ft) since the last ice age approximately 21,000 years ago.

Think for a second about what is even meant by *sea level.* First of all, the seas are not even close to level and certainly not flat to within a few millimeters. There are continuous waves, swelling of the ocean surface, and periods of calm and storm (Figure 3-3).

How can scientists measure a difference of a few millimeters a year? The noisy background superimposed on the surface of the sea, that is, waves, swelling, and calms and storms, must be factored out of the measurement, along with the tides. We are looking for changes on the order of a millimeter each year. By taking a large enough statistical sample, through, the trend pictured in Figure 3-4 emerges. Global sea level rose at an average rate of 1.8 mm per year between 1961 and 2003 (just a tad faster than the rate since 1800).

Figure 3-3 How do we define sea level? (*Source: NOAA.*)

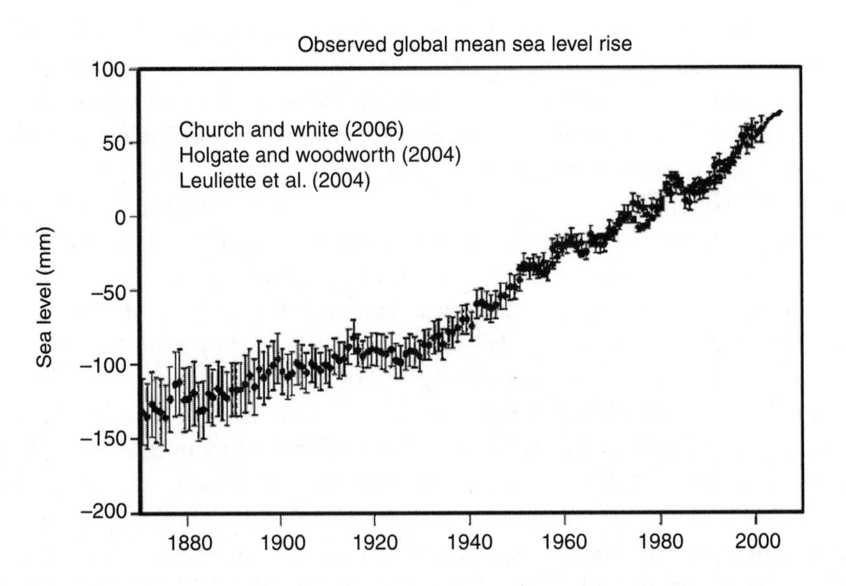

Figure 3-4 Global sea level rise from gauge and satellite measurements. Satellite altimeter data are shown by the solid dark line beginning in 1993. Global sea level rose at an average rate of 1.8 mm per year between 1961 and 2003. (*Source: IPCC.*)

Measuring Sea Level from Orbit

Satellite observation beginning in the early 1990s provided more accurate measurements than before and provided nearly global coverage. Satellites use a technique called *radar altimetry* that determines the distance between the satellite and the surface below based on the time it takes for a radar signal to traverse that distance (see Figure 3-5).

The Ocean TOPography Experiment (TOPEX) *Poseidon* satellite, a cooperative effort by the United States and France, had been in operation from 1992 until 2006. A successor (TOPEX *Jason*) went into operation in 2001. These satellites have been providing global data on sea level and the complex patterns that form in the open ocean covering 95 percent of the earth's waters.

The TOPEX satellites make an absolute distance measurement with respect to the center of the earth. By doing so, they are able to avoid difficulties associated with making measurements in relation to a fixed point on a nearby coastline. Sea level measurements made by the TOPEX *Poseidon* and *Jason* spacecraft are shown in Figure 3-5. Results are similar to those shown in Figure 3-4 but reveal a higher annual sea level increase because the satellite data include only the most recent data.

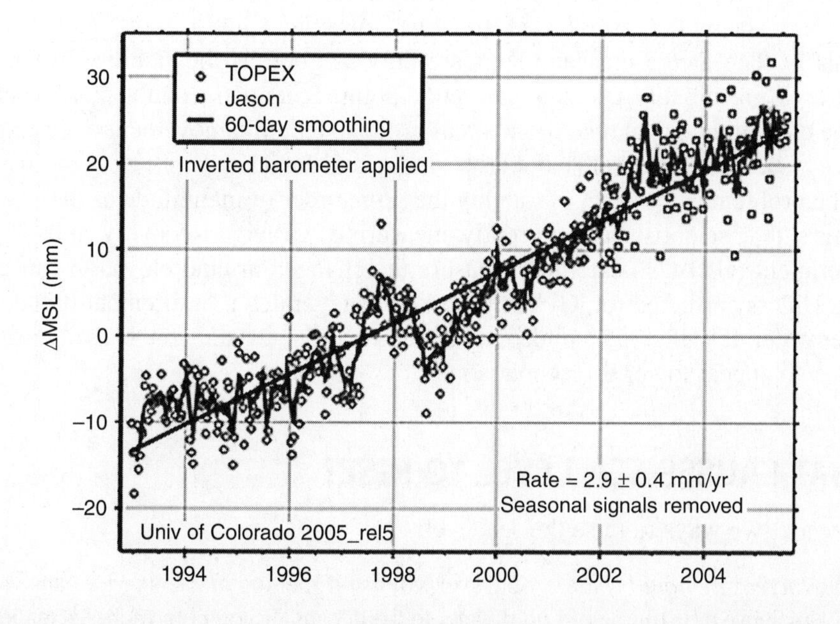

Figure 3-5 Sea level measured by TOPEX *Poseidon* and *Jason* satellites between 1994 and 2004. The increase in average global sea level is nearly 3 mm per year. Measuring sea level by satellite altimeters provides a way to track events such as the El Niño southern oscillation (ENSO).

These data show a significantly higher rate of sea level rise than in the previous half century and are confirmed by coastal gauge measurements. Satellites since 1993 show that sea level increase varies from region to region. The largest rise was found in the western Pacific and eastern Indian Oceans. Nearly all the Atlantic Ocean shows a sea level rise over the past decade. Sea levels in the eastern Pacific and western Indian Ocean have been falling. Some changes are influenced by atmospheric patterns such as the El Niño southern oscillation (ENSO) and the North Atlantic oscillation (NAO), which will be described in Chapter 4.

Separating the Movements of Earth and Oceans

The earth seems very solid to us. But when the earth supports a massive weight such as the polar ice sheets and glaciers, it compresses a little bit like silly putty. As the ice melts and the weight is removed, the earth springs back. This can take thousands of years and continues long after the ice is gone completely. This process is known as *global isostatic adjustment* (GIA). The last ice age occurred 21,000 years ago, and the earth is still "snapping back" to the shape it had before it bore the mass of the glaciers.

It is necessary for scientists to isolate possible vertical movements of land. Tide gauges measure sea level. These measure the average sea level with respect to a reference point on the land. As with temperature measurements, anomalies—or departures from the established baseline—are what are tracked and reported primarily. The gauge measurements are only as good as the stability of the land used as a presumably constant reference point. Scientists cannot make accurate gauge measurements along coastal areas prone to tectonic movement or earthquake activity, such as are found near Alaska, India, and Japan.

The problem is that GIA is roughly the same order of magnitude as the sea level changes that scientists are currently measuring. Gravity Recovery and Climate Experiment (GRACE) satellites measure the changes in land elevation caused by GIA. Unless corrected for, GIA will give an exaggerated measurement if occurring under water. It will give an understated reading if the coastal area is rising, causing the sea to appear lower by comparison.

WHAT CAUSES SEA LEVEL TO RISE?

There are two ways to raise the sea level:

1. *Increasing liquid volume.* Sea level will rise if melting glaciers or ice caps increase the volume of liquid water to the oceans. If you pour more ice tea into a glass, it will go to a higher level. This accounts for about half the rise in sea level.
2. *Thermal expansion.* The oceans are like a huge thermometer with all the liquid in the bulb. When the temperature of the oceans increases, the liquid

water—like just about any other material—will expand. This also results in a larger volume, which is responsible for roughly the other half of the measured sea level increase.

BACK-OF-THE-ENVELOPE CALCULATION: GLOBAL SEA LEVEL INCREASE FROM THEMAL EXPANSION

Determine the sea level increase resulting from the following (highly simplified) assumptions:

- The earth's radius is 6370 km.
- Water expands by a fraction that is 0.0021 of its original volume for each degree Celsius.
- The oceans are heated by 0.1°C (uniformly) to a depth of 0.7 km.
- The oceans occupy 70 percent of the earth's surface.
- The surface of a sphere = $4\pi r^2$.
- $\pi = 3.14$.
- 1 km = 1,000,000 mm.

Solution

The surface area of the earth's oceans = $(70\%) \times 4\pi(6370)^2 = 350{,}000{,}000\ \text{km}^2$.

The volume of earth's oceans absorbing heat = $0.7\ \text{km} \times 350{,}000{,}000\ \text{km}^2 = 250{,}000{,}000\ \text{km}^3$.

The oceans expand $250{,}000{,}000\ \text{km}^3 \times 0.00021 \times 0.1°\text{C} = 5250\ \text{km}^3$.

The height increase = volume expanded/area = 14.6 mm.

At a rate of 1.8 mm/year, during the 42-year time period, global sea level rose 76 mm.

Comment: Thermal expansion accounts for about one-quarter of the sea level rise. This simplified calculation shows that sea level rise was consistent with the contribution expected for this time period.

Melting Ice and Snow

Ice and snow that have been around for thousands of years have rapidly begun to melt and disappear. Approximately 10 percent of the earth's land and 7 percent or its oceans are permanently covered by ice. The massive transformation of much of this ice and snow to liquid water is probably the most conspicuous indication of global warming.

Winter comes later to the northern hemisphere and leaves sooner each year. Rivers and lakes freeze 5.8 days later and thaw 6.5 days earlier each year than they did a century ago. For every year since 1966, snow cover in the northern hemisphere has decreased by 5 percent.

ARCHIMEDES' PRINCIPLE—THE DIFFERENCE BETWEEN FLOATING ICE AND LAND ICE

If you place an ice cube in a glass of water and come back a few hours after the ice has melted, you will find that the level of the liquid has remained just where is was before. (We are assuming that there is no significant evaporation of water). The ancient Greek scientist Archimedes explained why this happens. Ice floats because the liquid it is floating in exerts a buoyant force that equals the weight of the floating ice. Archimedes discovered that the buoyant force on an object equals the weight of the water that is occupied by the part of the floating object that is underwater. This exactly equals the weight of the floating ice. As a result, when the ice melts, the same exact volume of water is returned to the sea as was occupied by the floating ice (Figure 3-6).

Arctic ice caps, icebergs, and ice shelves are like that ice cube. They float on seawater and do not contribute to an increasing sea level when they melt. Ice supported on land is another story. Glaciers, snow caps on mountains, and the Greenland and Antarctic ice packs will all add to the volume of the sea as they melt and contribute to a rise in sea level.

ARCTIC SEA ICE IN SUMMER

There is 7.4 percent less Arctic ice now during the summer months than there was 10 years ago, leaving a diminishing extent of permanent ice in that region. The melting of Arctic ice is more pervasive in the summer months than year round, and there is an annual 2.7 percent reduction in overall ice cover.

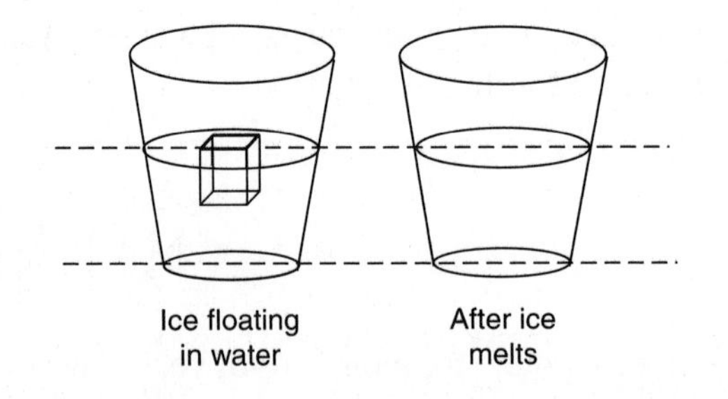

Figure 3-6 The melting of floating ice such as the Arctic ice caps does not contribute to sea level rise. (*Drawing by Ally Silver.*)

During the summer of 2007, spacecraft operated by the European Space Agency produced imagery that showed a brief opening in the Northwest passage through the Arctic Ocean. The amount of Arctic ice coverage that summer had shrunk to the lowest level on record. A sheet of ice the size of California that had been present since at least the 1970s when satellite imaging began had completely melted. A sea channel through the Arctic ice was once actually a dream of traders in the northern latitudes early in Western exploration and is actually listed by some people as one of the few positives that may come from global warming. However, this is not a positive for Arctic wildlife or for the indigenous human populations inhabiting the polar regions, whose traditional way of life is now threatened.

Total breakup of the Arctic sea ice could seriously compromise the ability of some Arctic animal species to survive. Seals bear their young and rest on the ice. Larvae of small fish are sheltered in the ice structure. These larvae develop into fish that are a source of food for beluga whales and narwhals. Loss of the Arctic ice sheets owing to global warming likely will have a significant impact on the fish and animals that depend on it for their survival.

Estimates put the Arctic polar bear population at between 20,000 and 25,000. Polar bears hunt and migrate on the Arctic ice sheets, frequently moving from one iceberg to the next in search of prey. It is not unusual for polar bears to trek thousands of miles across the ice. Some bears tracked by satellite travel as far as 150 miles across the open ocean. Polar bears are dependent upon seals that breed on the Arctic sea ice. Roughly one of twenty attempts to secure food is successful for the polar bear. As the sea ice coverage during the summer months becomes progressively smaller, the bears lose access to hunting areas.

Female polar bears are more severely affected than males. After their typical 5- to 7-month annual winter fast while nursing, females are in an especially weakened condition and emerge badly in need of nourishment. With Arctic ice shelves now breaking up 3 weeks earlier than they did in the early 1970s, the bears must come ashore earlier and, as a result, fast longer. With the bears' ability to capture food severely compromised, their prospects for survival are becoming increasingly difficult. In 1987, in one Arctic region there was an estimated polar bear population of around 1200 bears. By 2004, there were less than 950 bears. It is thought that similar trends can be observed throughout the Arctic. Scientists are now saying that if Arctic temperatures continue to rise between 2.5 and 3°C (4.5 and 5.4°F) above preindustrial levels, the risk of extinction of Arctic polar bears will become significant (Figure 3-7).

Most of the warming that has occurred over the past few decades has taken place in the northern hemisphere. Arctic sea ice has been declining at a rate of 9 percent every 10 years. Satellites map the ice cover of various regions of the earth using microwave detecting instruments. These satellites do not require sunlight and are capable of penetrating the clouds that often obscure the Arctic region. Compare the

Figure 3-7 Polar bears rely on Arctic ice to survive.
(Photo courtesy of Joel Garlich-Miller, U.S. Fish and Wildlife Service.)

images collected by the *Nimbus 7* and Defense Meteorological Satellite Program (DMSP) satellites in 1979 and 2003 (Figures 3-8 and 3-9).

The marked differences between the two images suggest that a significant climate change is taking place in the Arctic. National Aeronautics and Space Administration (NASA) scientists attribute the decrease in the extent of Arctic summer ice to the increase in atmospheric temperature, which has been increasing recently at a rate of 1.2°C during the summer over Arctic ice every 10 years. (The global average air temperature did not even increase by this amount in the past 100 years).

Geologists believe that there are vast untapped reserves of oil located on the sea floor of the Arctic Ocean. Now that the prospect of ice-free sea lanes in the Arctic may become a reality owing to global warming, countries adjacent to this resource have begun to establish their positions to exploit it. Scientists from the U.S. National

Figure 3-8 Minimum (summer) Arctic sea ice in 1979. (*Source: NASA.*)

Figure 3-9 Minimum (summer) Arctic sea ice in 2003. (*Source: NASA.*)

Snow and Ice Data Center (NSIDC) estimate that at this rate of melting, the Arctic Ocean will have no ice at the end of the summer of 2060.

GREENLAND

Greenland is covered by a huge central ice sheet with glaciers, which are like giant rivers of slush flowing to the surrounding sea. The largest of Greenland's glaciers is called Jakobshavn, and it has doubled its speed of descent to the ocean to roughly 37 m (120 ft) per day. It alone contributes 11 cubic miles of ice each year to the sea. Together, all the glaciers of Greenland contribute to the 54 cubic miles of ice that fill the surrounding fiords with melting icebergs. This is twice as much as was discharged into the ocean 10 years ago. The pace is quickening and has exceeded the expectations of many researchers studying the region.

As Greenland's glaciers progress seaward, the sections projecting toward and floating on the sea, called the *tongue*, break off and float away. However, Jakobshavn's tongue has receded 4 miles inland. Researchers are finding that when glacial tongues actually floated on the sea, they served the purpose of shoring up the glaciers and slowing their decent to the sea. With the tongues gone, the glaciers flow unimpeded at a faster rate than scientists previously thought possible. Satellite measurements such as those from the GRACE satellites, which monitor total mass in their field of view by detecting minute variations in the earth's local gravitational field, confirm Greenland's pace of melting.

Parts of Greenland actually appear to be cooling, with snow being added to the ice pack in the central part of Greenland, but melting is occurring along the coastal regions. Overall, there is more melting than buildup of new ice pack. The data plotted in Figure 3-10, collected by the GRACE satellites, show the overall ice mass of Greenland between 2002 and 2005.

The GRACE satellites are a pair of satellites orbiting together in tandem. They make extremely accurate measurements of the earth's gravitational field. These satellites do this by detecting small changes in the gravitational field by registering minute disturbances to their orbit. This enables the satellites to pick up very small changes in the mass directly below their orbital path. Reduced gravitational attraction means reduced mass, which is a sign that an ice sheet is melting.

The graph indicates that Greenland is losing ice mass at a rate of about 162 km^3 per year. Notice the periodic seasonal increases during the winter months and declines in the summer months and the overall downward trend. This amount of melting contributes an increase of 0.4 mm per year (0.15 in) to the sea level.

Snow melts seasonally as the warmer months approach. One indicator of the onset of a warmer climate is that the number of days that melting can occur increases. DMSP satellites measure ice and can detect the how many days ice melts during the summer months. These satellites indicated that snow melt occurred more than

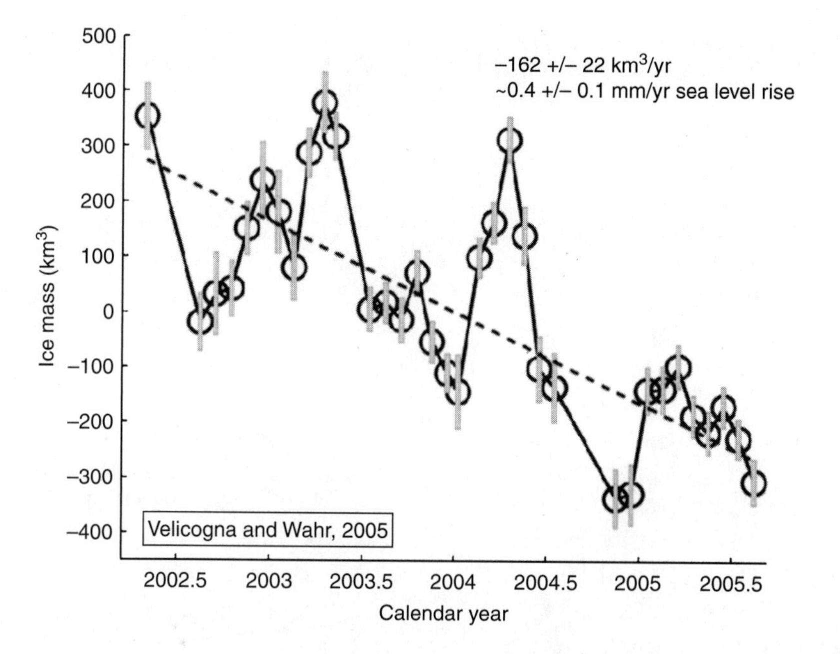

Figure 3-10 Ice mass of Greenland measured by the GRACE satellites. (*Source: NASA; credit: Velicogna and Wahr, 2005.*)

10 days longer in 2006 than in previous years. The darker regions of Figure 3-11 show the locations where melting occurred 5–10 days more in 2006 than the average number of days between 1988 and 2005.

If global warming continues at its current level, within as few as 50 years, the temperature threshold necessary to melt all of Greenland may have been crossed. Before we rush to redefine a new boundary for "ocean-front property," though, it is expected to take a few thousand years for all of Greenland to melt.

If all the ice and snow in Greenland melted, the seas would rise by about 7.3 m. A source of concern is that melting ice reflects only 50 to 60 percent of the sunlight striking it in contrast to fresh, dry snow, which reflects 80 to 90 percent of the incoming sunlight. Pools of liquid water forming on melting ice can be observed in satellite imagery such as Figure 3-12.

The more days melting occurs, the more additional heat is absorbed by the ice, further prolonging the time of melting. As meltwater flows across the ice and drains to the bedrock, it lubricates the ice, which then slides faster. Figure 3-13 shows a vertical shaft called a *moulin* filled with a stream of water generated by melting ice. This gives an idea of the powerful forces that can be set in motion when glacial melting is initiated. Once the melting process is set in motion, it can become self-perpetuating and accelerate the rate of ice loss.

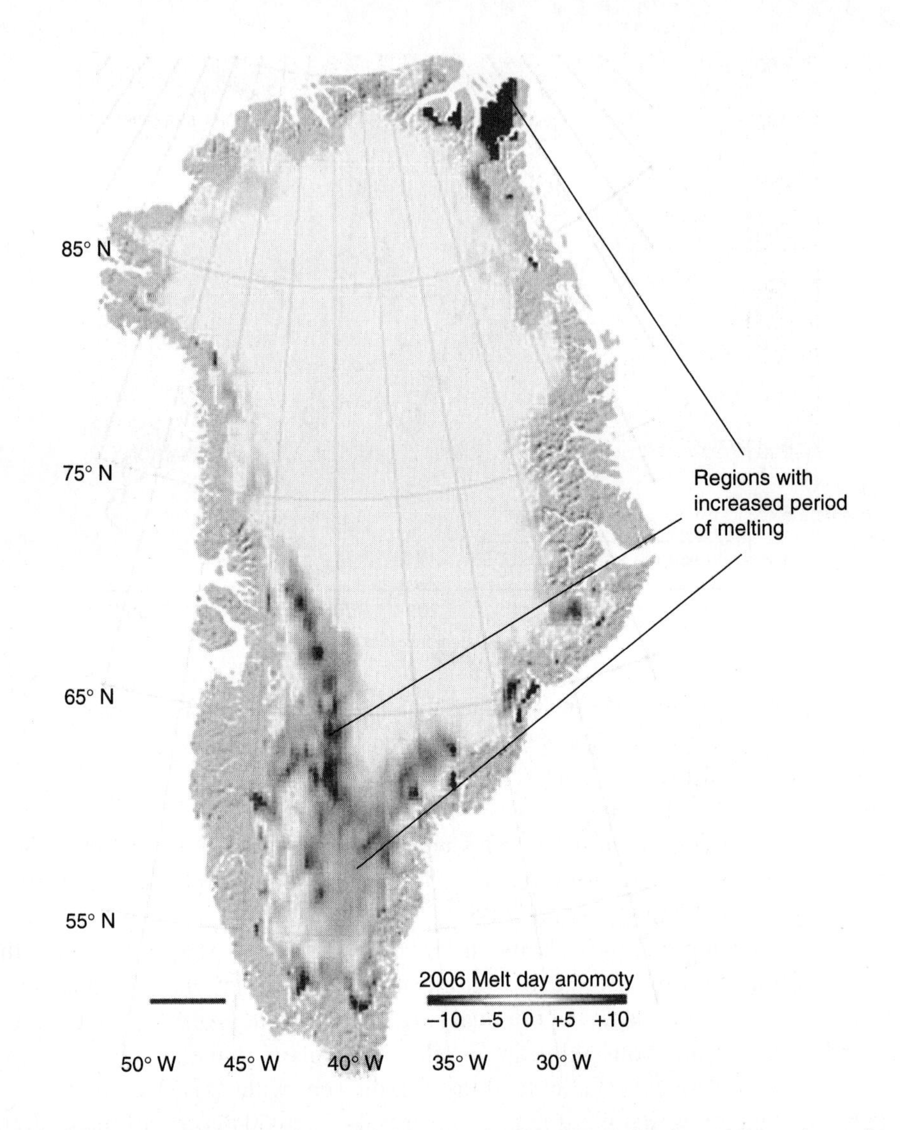

Figure 3-11 Number of days melting occurs in Greenland. Darker regions show places where average melting is more than 5–10 days above average. (*Source: NASA.*)

ANTARCTIC

Antarctica is simultaneously experiencing regions that are losing and regions that are gaining ice. Sections of the continent are partially isolated from the impact of global warming because they are at a higher elevation than the Arctic. The eastern part of Antarctica is growing—primarily as a result of increased snowfall—whereas the western regions of Antarctica are losing mass. Combining the two regions

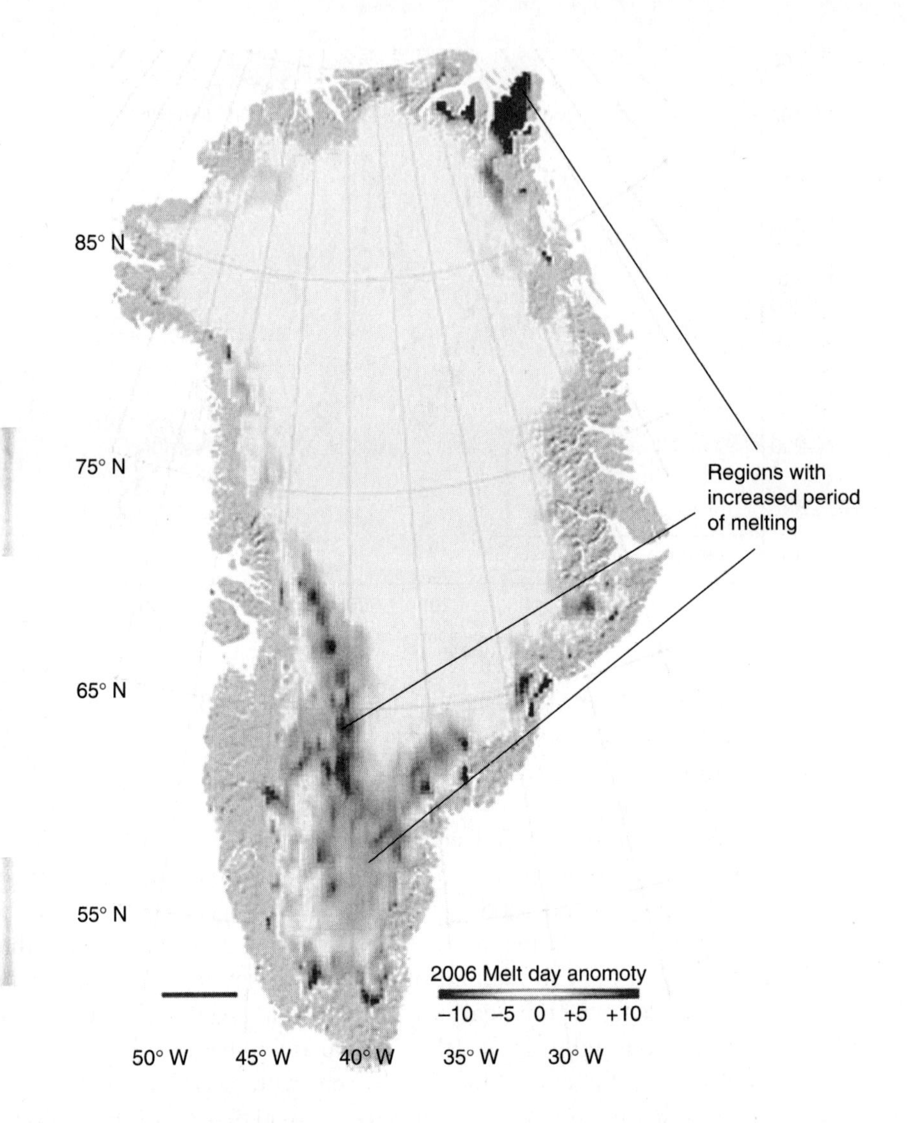

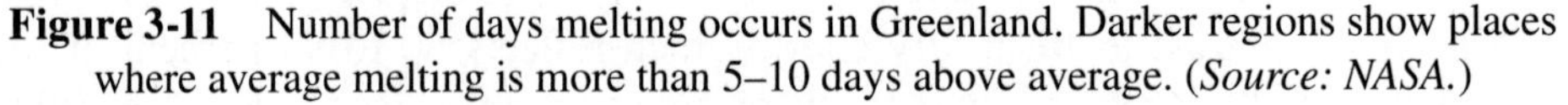

Figure 3-12 As ice melts, it absorbs more sunlight, causing accelerated melting. (*Source: NASA.*)

Figure 3-13 Meltwater stream from the Greenland ice sheet. (*Source: NASA.*)

results in a net loss of ice for the region. Estimates of overall ice sheet mass decline range from 50 to 200 billion tons per year.

Scientists had been watching the large ice shelves in the western part of Antarctica via satellites such as the Terra spacecraft, Landsat, and ICESat. One ice shelf section disintegrated in 1995. Usually, ice sheets project over the ocean until they become unstable, after which they break off and float away in a process that typically takes decades. Scientists tracked an ice shelf designated as *Larsen B* for several years as it receded from its previous positions (Figure 3-14).

Approximately 11 percent of the Antarctic continent is a floating ice sheet that does not contribute to sea level increase. The loss of the floating ice sheet is significant, however, because it frees up adjacent ice supported by land. The land-based ice will contribute to sea level increases as it melts. A glacier may appear to be a solid sheet of ice, but glaciers flow downhill under the force of gravity once the ice is above a minimal threshold beginning at around 18 m (60 ft) thick. Ice sheets can impede this glacial movement. However, if the ice sheets are removed, the glacier picks up speed as if someone just removed the emergency brake on a car parked facing downhill on a street in San Francisco. Scientists using the European Space Agency's ERS remote sensing satellite and Canada's RADARSAT found that glaciers flowed as much as eight times faster after the

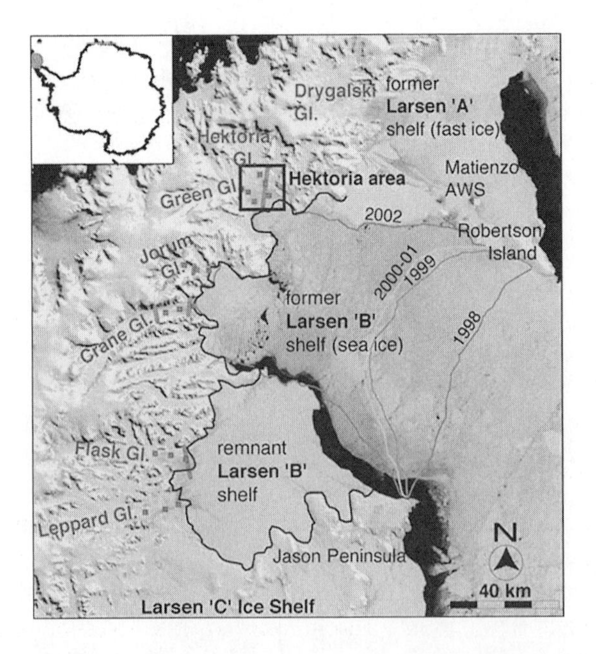

Figure 3-14 Larsen B ice sheet in Antarctica. Lines show the retreat of the ice shelf during the previous 6 years. (*Source: NASA; credit: NSIDC.*)

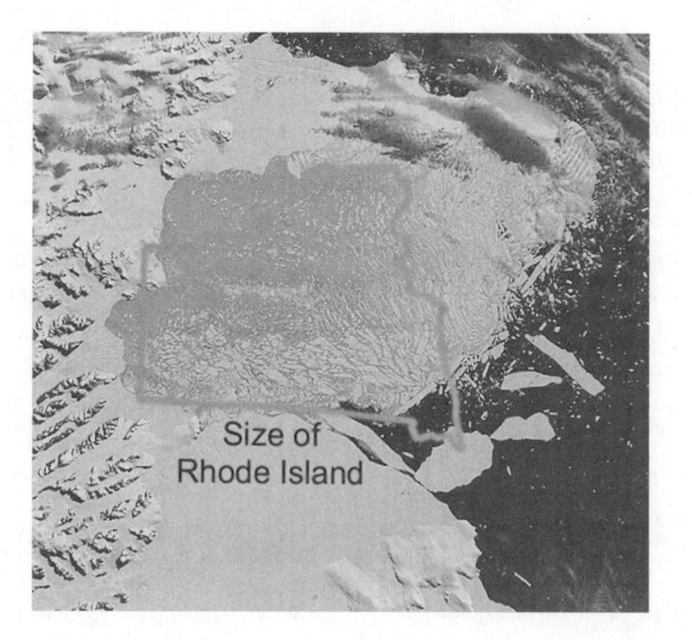

Figure 3-15 Larsen B after breakup. An area the size of Rhode Island disintegrated. *(Based on NASA Terra/MODIS imagery.)*

Larsen B ice shelf fell apart. Scientists were able to determine the speed of these glaciers by tracking the movement of holes and other markings on the surfaces of the glaciers.

Between January 31 and March 7, 2002, scientists watched Larsen B disintegrate. Figure 3-15 shows a section of ice the size of Rhode Island that fell apart in 35 days.

Air temperatures have decreased over parts of Antarctica during the past several decades. However, in the Antarctic Peninsula, where the Larsen B ice shelf is found, temperatures increased by 2.5°C (4.5°F) during that same time period. This is five times the global average temperature increase.

GANGOTRI GLACIER—THE HIMALAYAS

The Gangotri Glacier is one of the largest in the Himalayan mountain range. It has been slowly receding since 1780, but like many other glaciers around the world, it has been melting much more rapidly since the 1970s. The glacier is now about 30.2 km (18.8 miles) long and 0.5–2.5 km (0.3–1.6 miles) wide. Over the past 25 years, the Gangotri Glacier (shown in Figure 3-16) retreated more than 850 m (2788 ft). During the 3 years between 1996 and 1999, the glacier lost 76 m (249 ft).

Figure 3-16 Gangotri Glacier retreating rapidly in recent years. (*Source: NASA.*)

The glaciers surrounding nearby Mount Everest have decreased by 2–5 km (1–3 miles) over the past 50 years. The runoff has contributed to flooding of glacial lakes and surrounding communities. As glaciers melt, greater pressure is placed on natural ice or moraine dams. When these dams fail, neighboring communities can be exposed to *glacial lake outburst floods*. Regions that are at highest risk include Central Asia, the Andes regions of South America, and the Alps in Europe. The Tsho Rolpa glacial lake in Nepal is growing and is considered among the most vulnerable. From 1940 to 1970 there was roughly one glacial lake outburst flood per decade. In the 1990s, this has increased to roughly one in every three years.

ALASKA—MUIR GLACIER

Alaska is warming more rapidly than most other places on the earth. Naturalist John Muir observed the Muir Glacier in Alaska's Glacier Bay National Park and Preserve in 1879 while on an expedition. Notice how much this glacier has retreated between August 13, 1941, when the first picture was taken, and the more recent shot in 2004 (Figures 3-17a and 3-17b).

By 2004, ocean water replaced the glacier in the valley. The end of the glacier, which at one time was 200 ft (61 m) high, has now receded so far that it is out of the field of view of the picture. Vegetation now covers what was once bare rock in the foreground.

Figure 3-17a Muir Glacier 1941. (*Source: NSIDC; credit: W. Field.*)

Figure 3-17b Muir Glacier 2004. (*Source: NSIDC; credit: B. Molina.*)

CHACALTAYA GLACIER—BOLIVA

The Chacaltaya Glacier in Bolivia stands 17,250 ft above sea level. It was considered to be the highest ski slope in the world. South American skiing championships took place on the half-mile run. The glacier was already slowly shrinking when the ski area first opened in 1939. But the melting has accelerated rapidly to the point where there is not much left today except for a few patches of rocky ice. This progression is shown in Figure 3-18. The glacier lost two-thirds of its mass since the 1990s and may disappear entirely by 2010.

Figure 3-18 Chacaltaya Glacier. (*Source: IPCC.*)

PERMAFROST

Tundra is found near the poles and at high altitudes and is characterized by perpetually frozen ground called *permafrost*. The severe frigid conditions of the tundra inhibit the growth of trees, leaving cold-tolerant vegetation such as moss, heath, and lichens to dominate the landscape. Permafrost covers roughly 20 to 25 percent of the earth's land areas and serves as a sensitive indicator of climate change. In some places such as northern Siberia, the permafrost can be 1.6 km (about 1 mile) thick. Air temperatures above the permafrost layers of the Arctic have increased by 3°C since the 1980s. Recent thaw rates are typically about 2–4 cm (0.8–1.6 in) each year in many places around the world. Melting of permafrost is another sign of global warming.

In South-Central Alaska, the Copper River and Northwestern Railway was constructed across what was at the time considered to be a permafrost-covered region of that state. The railroad was built to provide passage across difficult terrain. This region warmed as global warming came to Alaska. Thawing of the permafrost bed produced the roller-coaster-like waviness in the track shown in Figure 3-19. Although officials abandoned this section of track in 1938, much of the disruption was the result of accelerated thawing in recent years.

Tibet is an isolated alpine region containing more that 17,000 glaciers. Travel between China and Tibet has been a long and challenging ordeal, often taking

Figure 3-19 Copper River and Northwestern Railway track built on Alaska permafrost. *(Source: NASA; credit: United State Geologic Survey.)*

6 months to a year. In 2007, China plans to construct a railway connecting its northern provinces and Tibet. Half the 1118 km (695 miles) of track will be built on permafrost. When the project was first planned, scientists expected that the region's air temperature would rise by 1°C in about 50 years. The belief at present is that as a result of global warming in the region, air temperature will increase by between 2.2 and 2.6°C (4 and 4.7°F). Thus the builders of this railroad, having learned the lesson of the Copper River and Northwestern Railway, are using engineering techniques to enable them to adapt to the climate changes that they anticipate for that region. Engineers are rerouting some sections to avoid the most unstable permafrost areas, erecting overpasses across questionable terrain, and building an insulating layer to protect the frozen ground.

Increasingly, as global warming occurs, people in various parts of the world will need to adapt to the new climate conditions that are produced. With improved monitoring by satellites, scientists can track changes in permafrost conditions over a wide part of the earth and with much greater frequency. This can have a direct bearing in areas where infrastructure is at risk and can provide a critical indication of the extent of climate change.

Impact on Wildlife—Some Examples

CORAL

Coral reefs are the skeletons of tiny marine animals that serve as habitats for about one-fourth of the world's marine species. The individual coral polyps typically form colonies that establish a symbiotic relationship with algae. The algae contribute some of the color characteristic of a living reef. In takes years for the corals to secrete enough calcium carbonate to form a reef.

Changes in ocean temperature or salinity are believed to lead to a condition affecting coral reefs called *coral bleaching*. As shown in Figure 3-20, parts of the coral turn white, indicating loss of the algae.

Coral bleaching has become more common since the 1970s as a result of increasing ocean temperatures associated with global warming. IPCC scientists estimate that as much as 80 percent of the coral reefs in the Caribbean may have been lost. The 1998 El Nino event killed robust coral colonies that may have persisted for centuries. It is unlikely that coral will be able to adapt to increased temperature by migrating. However, scientists are watching to determine whether species of algae that are tolerant of a slightly higher temperature may delay the complete loss of coral reefs.

Figure 3-20 Coral showing areas of bleaching. (*Source: NOAA.*)

GOLDEN TOAD

Most people probably have never seen or even heard of the golden toad. This creature (shown in Figure 3-21) once occupied a very narrow ecologic niche that included a habitat in a 5-km^2 region of Costa Rica. The golden toad is now believed to be extinct.

These toads used to lay their eggs during the spring months in pools of rainwater. Warmer temperatures resulted in decreased local rainfall. In 1987, 1500 toads were observed. Since 1989, no toads have been found despite extensive searching. This species represents a potential loss of biodiversity worldwide. Some of the early effects of this loss of biodiversity are becoming noticeable around the world, but loss of the golden toad serves as a warning of possible things to come.

MOUNTAIN PINE BEETLE

The mountain pine beetle inhabits the forests of the western part of North America. Mountain pine beetles infest trees such as spruce and ponderosa pine. These beetles bore through the bark of the trees, resulting in their destruction. The spread of the beetles is limited by frost. With the number of frost-free days increasing, the pine beetles have that much more time to damage trees. As a result, many additional acres of evergreen forest in the Pacific Northwest have been destroyed.

Figure 3-21 Golden toad. (*Source: U.S. Fish and Wildlife Service.*)

Wacky Weather

TROPICAL CYCLONES, HURRICANES, AND TYPHOONS

Sea surface temperatures greater than 26°C (79°F) are needed to produce hurricane-force winds. Since hurricanes derive their energy from the ocean's heat, greater sea surface temperatures can be expected to contribute to the frequency and severity of hurricanes. Records of the destructiveness of hurricanes show a significant increase since the 1970s. The number of named storms since 1994 has risen. Overall, storms last longer and are more intense. This correlates with the observed increase in the number of hurricanes that reach category 4 or 5 status since 1970.

This was the case in the summer of 2005. Because of a prevailing El Nino, the sea was warmer than usual that year. The sea, especially in tropical areas where there is little reflection from the water and where the sun's ray are most direct, absorbs solar heat very efficiently. Hurricane Katrina was unusually powerful, and its effects were especially severe because of its impact on an especially vulnerable coastal area (Figure 3-22).

Katrina dropped about 30 cm (12 in) of rain on the Gulf Coast states of the United States in 2005. One estimate (by Kevin Trenbeth. *Warmer Oceans, Stronger Hurricanes, Scientific American,* July 2007) claims that 8 percent of this storm's

Figure 3-22 Hurricane Katrina about to make landfall, monitored by a weather satellite. (*Source: NOAA.*)

total rainfall, or 2.5 cm (1 in), was the direct result of global warming. However, global warming most likely did not cause Katrina. It is probably more accurate to say that the storm was intensified by perhaps one category level by the excess heat in the ocean at the time.

The next summer, many people expected that a new precedent had been set and braced for what they thought might be a repeat of the previous intense hurricane season. However, 2006, to the surprise of many forecasters, was a very calm year. This was not because global warming had suddenly stopped but rather because El Niña conditions were occurring that resulted in cooler sea surface temperatures that year.

In the summer of 2002, there were widespread floods in Europe. The next year, 2003, brought record-breaking heat waves and drought. Weather has it own drivers, including recurring natural climate cycles (such as the ENSO). It appears that global warming is making El Niño conditions more severe. Figure 3-23 shows a pattern of tropical storms based on data from the National Hurricane Center *(compiled by Neumann and coworkers).*

Notice that there is a recent increase for all storms, hurricanes, and major hurricanes. There are also long-term fluctuations in all three records suggestive of natural cycles such as the ENSO. For instance, there is a lull in the 1920s and an increase in intensity toward the end of the 1930s. In all fairness, some of the recent upswing in major hurricanes is partly a result of more thorough recent monitoring. In particular, the use of surveillance flights into the eyes of hurricanes has become

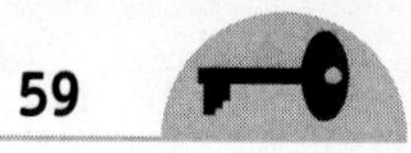

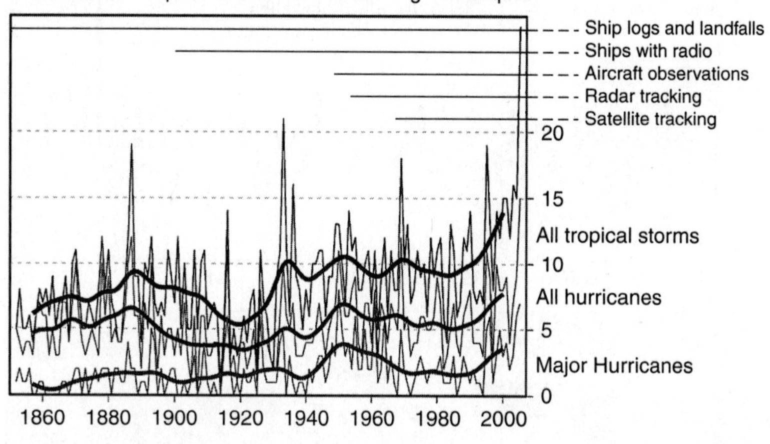

Figure 3-23 Tropical storms, hurricanes, and major hurricanes. (*Credit: Robert A. Rohde,Global Warming Art.*)

more routine in recent years and has made it possible to classify storms more accurately—particularly the most severe ones—on the Safir-Simpson scale which measures the intensity of hurricane strength.

There have been storms, droughts, and floods in the past. Tropical storms are strongly influenced by El Niño/ENSO, with an increase in one part of the world coinciding with a decrease in another part. This also has caused a movement of the jet streams closer to the poles (since the 1960s), resulting in conditions that favor the formation of tropical storms.

It is premature to definitively attribute current storm patterns to climate change without first separating natural climate cycle (e.g., ENSO, etc.) effects.

PRECIPITATION

Higher temperatures promote more evaporation. Overall, the total amount of global precipitation does not show an increase. However, there is a statistical increase in the incidence of heavy (95th percentile) and very heavy (99th percentile) precipitation events worldwide—in terms of extreme events and total annual rainfall. Figure 3-24 shows the trend toward the increase in the percentage of annual rainfall contributed by extreme amounts. This shows the change in the percentage of very wet days compared with a running baseline of 22.5 percent established from 1951–2003.

It is drier in the Sahel, the Mediterranean, and southern Asia. It is wetter in North America and western Europe. Precipitation has been increasing in the midlatitudes,

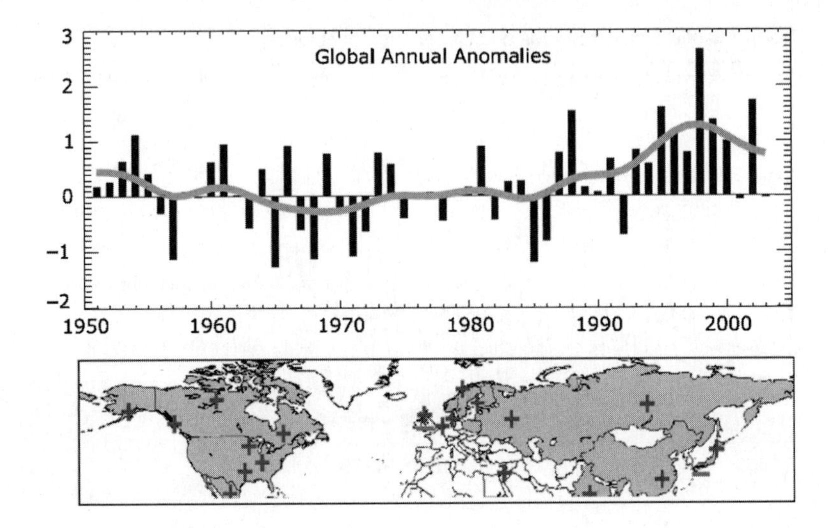

Figure 3-24 Wet days worldwide. Areas with increased precipiation are shown by + sign. (*Source: IPCC.*)

including eastern North and South America, northern Europe, and northern and central Asia. Figure 3-25 gives an overview of how some parts of the world are getting wetter and some drier.

For every 1°C rise in air temperature, the atmosphere can hold an additional 7 percent water vapor. An increase in air temperature over bodies of water can increase evaporation. This can create conditions that promote more precipitation. The higher atmospheric temperatures also can enhance evaporation from soil, leaving it prone to drought. Whether there are moist or dry conditions depends on other climate factors such as El Nino or the NAO.

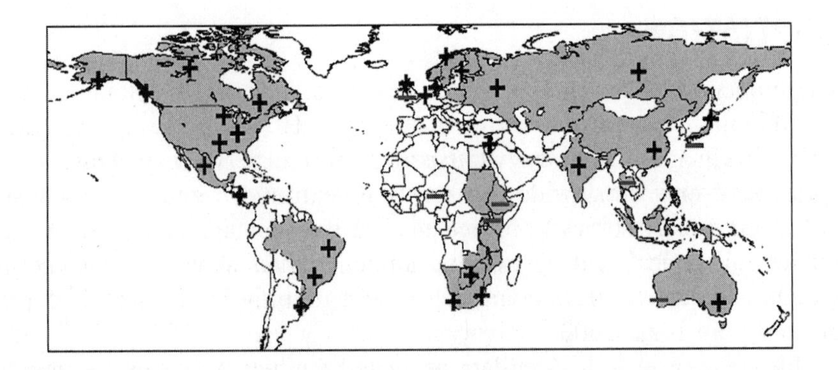

Figure 3-25 Distribution of wet days worldwide.

DETECTING CHANGES IN THE WEATHER

Measuring changes in weather conditions is in some ways similar to trying to determine whether the tide is coming in or going out. Suppose that we are sitting on the beach watching the waves. We may be very conscientious and watch closely for a half hour and conclude that the tide is not moving. We even may conclude that we do not believe in tides and think the whole thing is a myth. If we wait long enough, we will find that our beach blanket is soaked and our cooler is floating out to sea. Had we measured the furthest extent the water encroached on the sand and the furthest retreat back to sea, we would find a statistical correlation sooner than looking for a more dramatic overall movement. As with weather patterns, there are natural cycles of the spring and neap tides and the impact of local conditions such as storm. What may appear to be a change in climate actually may be a temporary condition caused by El Niño or other climate cycles. The onset of climate change is often found in statistical correlations rather than in more conspicuous changes to weather.

WATER VAPOR

Higher temperatures cause an increase in atmospheric water vapor, which, in turn, leads to even higher temperatures. Every 10 years, the atmosphere holds 1.2 percent more water than the previous decade. The relative humidity of the atmosphere has remained pretty much constant, but a greater amount of water is held in the atmosphere as sea temperatures increase. Water vapor is a natural greenhouse gas.

WARM DAYS

The year 2005 was one of the warmest on record. Eleven of the past 12 years (through 2006) were the warmest since 1850. Since 1950, the number of heat waves has increased. Probably the clearest weather-related trend is toward a decline in the number of coldest days (and nights) and an increase in the number of warmest days (and nights). There have been a greater average number of warm nights and a smaller average number of cool nights worldwide since 1951. This pattern is noticeable in Figure 3-26. The data show the number of extremely warm days worldwide, defined as being above the 90th percentile.

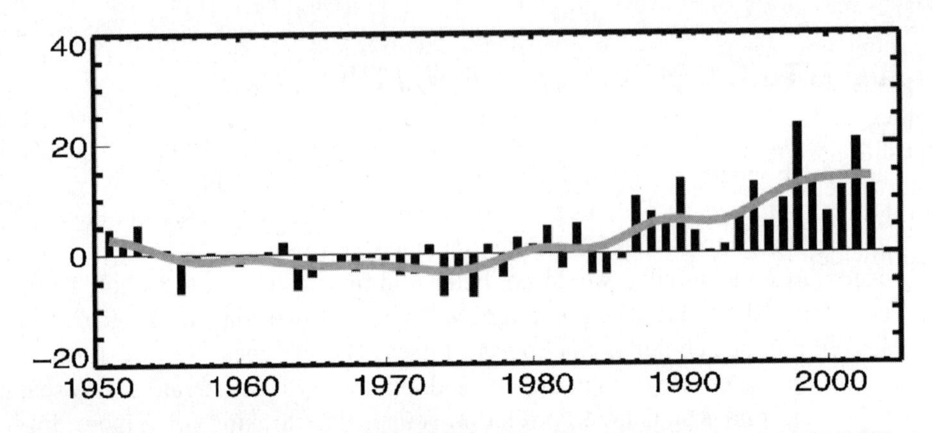

Figure 3-26 The number of warm nights has increased compared with 1951–2003. (*Source: IPCC.*)

DROUGHT

Drought has become more common, especially in the tropics, since 1970. Droughts are becoming longer and more severe and are affecting a wider area. This is consistent with decreased precipitation in some areas and higher temperatures causing drying conditions. Increasing sea surface temperatures and loss of snow are direct contributors to drought. An index called the *Palmer Drought Severity Index* (PDSI) is used to compare the extent of local loss of surface land moisture. Figure 3-27 shows that

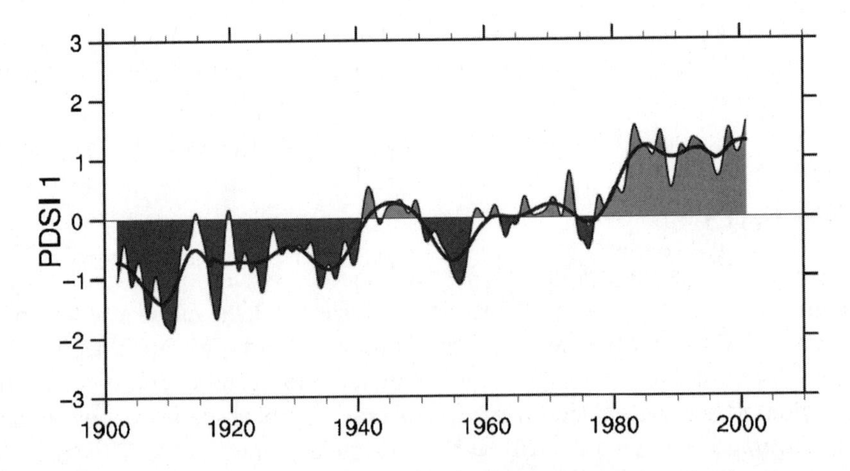

Figure 3-27 The Palmer Drought Severity Index (PSDI1) from 1900–2002 shows a tendency toward increasing drought conditions worldwide. (*Source: IPCC.*)

drought conditions overall have been increasing worldwide since 1900. Some regions, such as the southern part of South America, are not seeing drought but rather an increase in moisture. However, more areas than not are seeing drying conditions. The smooth line through the data shows a pattern of natural increases and decreases on the time scale of decades. The dominant pattern that emerges through the natural variations is a trend toward more severe drought conditions worldwide.

Key Ideas

- Average global sea surface temperatures increased by 0.35°C from 1961 to 2003.
- It takes more heat to change the temperature of a given volume of water than it does for an equal volume of air.
- The earth's oceans absorb 20 times as much heat as the atmosphere.
- The pH level in the earth's oceans has increased, indicating an increased absorption of carbon dioxide.
- Local salinity increases indicate freshening of seawater from glacier and ice-cap melting.
- The average global sea level has increased by 1.7 mms (0.07 in) per year since the 1800s (1.8 mm per year since 1961–2003).
- Recent satellite measurements indicate that the average global sea level is currently increasing at 3 mm (0.12 in) each year.
- The average global sea level has increased by 120 mm (4.7 in) since the last ice age.
- Sea level rises for two reasons: (a) water expands when it gets warmer, (b) water volume increases from meltwater. Each contributes in equal proportions to sea level increase.
- Arctic sea ice is 7.4 percent less in summer and 2.7% less overall than 10 years ago.
- If floating ice such as the Arctic ice cap melts, global sea level will not be affected.
- Rivers and lakes are freezing 5.8 days later and are thawing 6.5 days sooner each year than 100 years ago.

- Snow cover in the northern hemisphere is decreasing by about 5 percent each year.
- Albedo (percent of light reflected) for fresh snow is 80 to 90 percent and for melting ice is 50 to 60 percent.
- Greenland is showing 5–10 more melt days each year (than between 1988 and 2005).
- If all the snow and ice in Greenland melted, global sea level would increase 7.3 meters (24 feet)
- Despite breakup of some ice sheets in the western part of Antarctica, signs of massive melting have not been identified.

Review Questions

1. Which of the following is a significant cause of the sea level rise that occurred over the past century?
   (a) Permafrost melting
   (b) Thermal expansion of water
   (c) Increased precipitation
   (d) Melting icebergs
2. Where is most of the sun's energy absorbed when it hits the earth?
   (a) Oceans
   (b) Land
   (c) Air
   (d) Glaciers
3. What happens when carbon dioxide dissolves in the oceans?
   (a) Salt content is decreased.
   (b) The sea level rises.
   (c) There is more evaporation.
   (d) Seawater becomes more acidic.
4. What is the effect of Arctic sea ice melting?
   (a) Sea level rises.
   (b) There is no effect on sea level.

(c) Ocean salinity increases.

(d) Sea level falls.

5. If complete melting occurred, which of the following would cause the largest rise in sea level?

   (a) Antarctica

   (b) Greenland

   (c) Arctic sea ice

   (d) Permafrost

6. If both Greenland and Antarctica were to melt completely, approximately how much would the average global sea level rise as a result?

   (a) 0.64 m (2.1 ft)

   (b) 6.4 m (21 ft)

   (c) 64 m (210 ft)

   (d) 640 m (2100 ft)

7. Based on the most recent TOPEX measurements of sea level rise, how much higher would sea level be expected to rise in the next 10 years?

   (a) 0.3 mm (0.01 in)

   (b) 30 mm (1.2 in)

   (c) 300 mm (1 ft)

   (d) 3.0 m (9.8 ft)

8. How do the GRACE satellites monitor the mass of ice on the earth's surface?

   (a) Visible light

   (b) Microwave frequencies emitted by the earth

   (c) Reflected laser light

   (d) Local gravitational pull of the earth

9. If a scientist does not make corrections for global isostatic adjustments to the coastal landmass, how will sea level measurments be affected?

   (a) Sea level will appear higher than it is.

   (b) There would be no effect on the sea level measurement.

   (c) Sea level will appear lower than it is.

   (d) A false indication of sea surface temperatures will be given.

10. How would polar bears most likely be affected by loss of Arctic ice during summer months?
    (a) Minimal impact because they would be hibernating anyway.
    (b) They would find more prey because seals would become more abundant.
    (c) Their numbers most likely would decrease because of difficulties in finding prey.
    (d) They would find other animal species to prey on.

PART TWO

Why Climate Changes

CHAPTER 4

The Earth's Thermostat—Keeping the Earth Warm

Throughout its history, the earth has received heat from the sun. The earth reflects or releases a certain fraction of the sun's energy right back into space. This results in a thermal balance that has enabled the earth to remain a hospitable planet for most of its inhabitants most of the time. What is truly remarkable is how incredibly steady the average temperature of the earth has been for so long. As we saw in Chapter 3, a change in the average global air temperature of a few degrees is a big deal, and a swing of around 5°C (9°F)—up or down—puts us at the extremes of all-time historical records. We recognize that the earth is currently in a warming mode. This chapter will address what is driving that trend. There are many natural factors that influence the earth's temperature on a time scale that ranges from days to hundreds of thousands of years. We will discuss these natural cycles and the changes that have been set into motion since humans found a way to derive energy by burning fossil fuels. This chapter is about energy—how much the sun sends to us, how much the earth gets to keep, and how and why that balance is changing.

The Earth's Orbit

The earth goes through cycles of day and night and an annual progression of the seasons. The earth's axis of rotation at times points toward and then away from the sun. Throughout all this, the temperature of the earth averages out to establish a very stable historical baseline. So, when we measure a statistically significant change that approaches just 1°C, it gets scientists' attention.

If the earth's orbit were a perfect circle, its distance from the sun would not change. However, as is the case with all the planets in the solar system, the distance from the sun varies as the earth dutifully follows the precise elliptical path determined by the laws of Newton and Kepler. This means that at some points during that orbit, as pictured in Figure 4-1, the earth is slightly closer to the sun than at other times.

Because of the shape of the earth's path around the Sun, the earth is *closer* to the sun in the winter months. The earth actually receives *more* solar energy in the winter than in the summer. Figure 4-2 shows how the solar intensity varies throughout the year.*

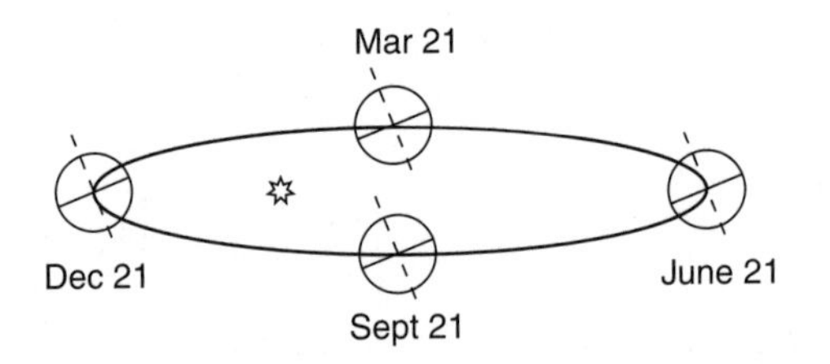

Figure 4-1 The earth's path around the sun is elliptical.

*Solar intensity varies inversely with the square of the distance between the earth and the sun. At its closest approach, the earth is about 147 million km from the sun and at furthest distance it is 152 million km away. Taking the square of the ratio of these distances, we find that solar intensity varies by about 6.9 percent between the least intense and most intense orbital positions.

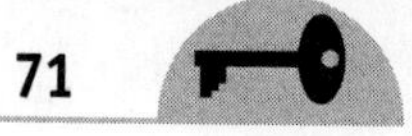

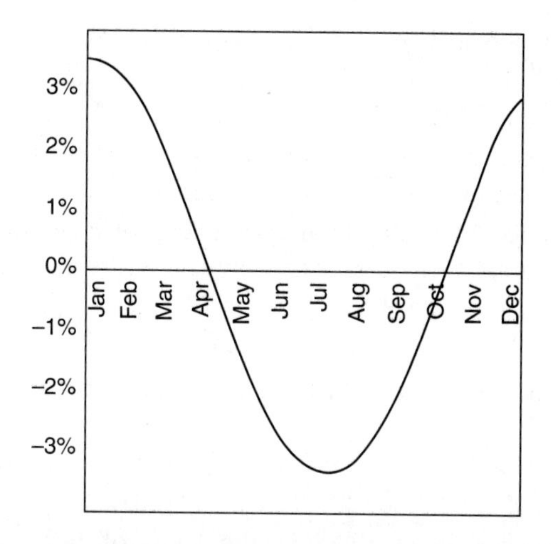

Figure 4-2 The sun's energy is most intense during the winter months in the northern hemisphere.

This may seem surprising at first because it is it is colder in the northern hemisphere during the winter months (December through February). However, the reason for the lower winter temperature is that in winter the sun strikes the northern hemisphere at a greater angle than during other seasons. The incoming sunlight is spread over a greater area, causing the northern hemisphere winter to be cooler despite the fact that the sun is closer to the earth. Since the earth spins like a barbequed chicken on a rotisserie as it goes through its daily day-and-night temperature swings, these variations do not affect the *average* temperature. The orbital variations repeat year to year. When all is said and done, the average annual temperature of the earth for many centuries has held within a tight range of only a few degrees.

How the Sun Warms the Earth

RADIATION—HOW THE SUN SENDS ENERGY TO THE EARTH

What Is Sunlight?

The sun tirelessly provides an average of about 1370 W of power (in the form electromagnet radiation to every square meter of surface it strikes. This amount of solar power is called the *solar constant* because is does not change except for minor variations (which will be addressed later in this chapter).

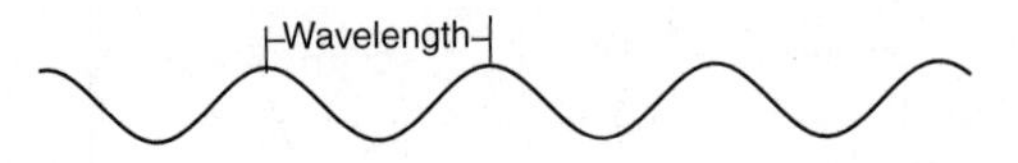

Figure 4-3 Wavelength—the difference between blue light and red light.

Energy from the sun is received in the form of light waves. Light is an *electromagnetic wave*. The frequency of the light—or how fast the waves vibrate—determines the different colors of the light. Blue light waves vibrate more vigorously than do red light waves. Since the waves of blue light vibrate faster, they are closer together, as shown in Figure 4-3. If the waves are closer together, they are said to have a shorter *wavelength*. Blue light has a shorter wavelength than red light.

The light hitting the earth from the sun is a mix of visible light and the invisible infrared and ultraviolet components of the spectrum. The mix of colors hitting the earth is shown in Figure 4-4. Most of the light energy coming to the earth from the

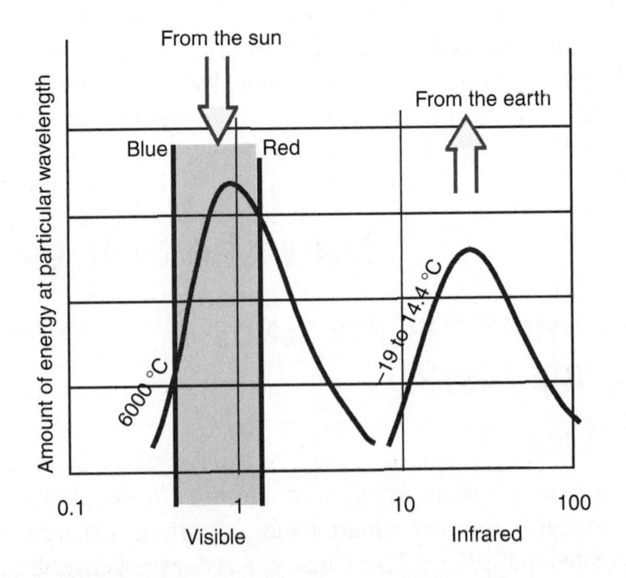

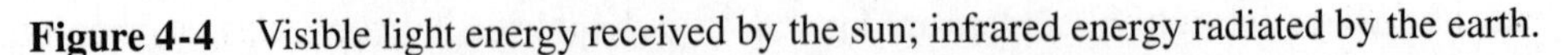

Figure 4-4 Visible light energy received by the sun; infrared energy radiated by the earth.

sun is in the middle of the spectrum, midway between blue and red. There is more green light than either red or blue light in the spectrum of light coming from the sun.

Since the sun's temperature is around 6000°C (10,800°F), it glows—not quite "white hot" but with a yellowish average color. The heating element of an electric stove, by comparison, might reach 800°C (1470°F) and have a reddish glow. As the earth is heated, it also glows in its own way. However, the earth is at a much colder temperature than the sun, namely, 14.4°C (58°F) at the surface to –19°C (–2°F) at the top of the atmosphere. For this reason, the earth's "glow" is *infrared.* (Joseph Fourier called the radiative energy coming from earth "dark heat.") This is *not* light reflected from the earth. Rather, it is light that is absorbed and then reradiated. All objects that are above absolute zero in temperature emit this type of radiation. The range of infrared wavelengths emitted by the earth is shown to the right in Figure 4-4. As we shall see shortly, the distribution of infrared radiation coming from the earth that is measured by satellites is not quite as smooth and uniform as this graph for a very important reason.

Living in a Greenhouse

The idea of a greenhouse is for light to pass through the glass. This light then is absorbed by the objects it strikes. The heated objects then reradiate infrared light, which cannot pass through glass.

In 1829, the French physicist and chemist Joseph Fourier developed a concept for how planets such as the earth maintain a steady temperature. He proposed that not only do planets receive energy from the sun, but they also radiate heat back into space. By suggesting that gases in the atmosphere increase the temperature of the earth, Fourier came up with the idea that became known as the *greenhouse effect.* Like a greenhouse, the earth receives energy from the sun. The atmosphere is like a clear pane of glass, and most of the light passes through it without being absorbed.

British physicist John Tyndall, working in the 1860s, studied absorption of light by different gases, including coal gas, carbon dioxide, and water vapor. Tyndall showed that visible light passes fairly well through carbon dioxide but that *infrared light* is very strongly absorbed. He saw this as possible cause of climate change and a possible explanation for the advance and retreat of glaciers. The two gases that make up most of our atmosphere, oxygen and nitrogen, do not absorb much light in either the visible or the infrared range. Carbon dioxide and other infrared-absorbing gases let visible light come through but trapped heat-producing infrared light.

In 1896, the Nobel Prize–winning Swedish chemist Svante Arrhenius (pictured in Figure 4-5) turned his attention to understanding what might have caused the ice ages that have come and gone throughout the earth's geologic past. Fossil records showed that ice covered the earth as far south as Germany and Illinois as recently as 12,000 years ago. Drawing on Tyndall's and Fourier's discoveries, Arrhenius

Figure 4-5 Svante Arrhenius calculated how carbon dioxide in the atmosphere affects the earth's temperature.

proposed that carbon dioxide released by ancient volcanoes resulted in the earth growing 20–30°C (68–86°F) warmer as a result of the greenhouse effect. He theorized that the decrease of carbon dioxide in the atmosphere between periods of volcanic activity resulted in the cooling periods that brought on the ice ages.

Arrhenius estimated that decreasing the amount of carbon dioxide by half (and taking into account the reduction of water vapor in the air at that lower temperature) would cause a drop in global temperatures of 4–5°C (7–9°F). Similarly, he predicted that doubling the carbon dioxide level would increase the earth's temperature by 5–6°C (9–11°C). By comparison with the more sophisticated climate models used today, Arrhenius' estimates are a little beyond the most pessimistic of today's projections. Today's estimates put the increase from a doubling of carbon dioxide levels at closer to 2–4.5°C (3.6–8.1°F).

Arrenhius was successful in identifying the relationship between carbon dioxide and global warming that is the basis of what is called the *greenhouse effect*. Although he was not correct in providing an explanation for the cause of the ice ages, he laid the foundation for developing a quantitative model to determine how a change in the concentration of carbon dioxide could affect the atmospheric temperature.

In 1938, Guy Stewart Callendar, a British coal engineer, analyzed temperature measurements taken from weather stations and concluded that the average temperature of the atmosphere was increasing. He attributed this rise in temperature to the buildup of carbon dioxide in the atmosphere as a result of burning fossil fuels.

What Happens to the Sun's Energy When It Gets to the Earth?

Figure 4-6 summarizes the various incoming and outgoing components of the sun's energy once it gets to the earth.

This is how it works:

1. Light comes from the sun. About 1370 W of power strikes every square meter of (perpendicular) surface. Half the earth is facing away from the sun at any given time. In addition, taking into account the curvature of the earth results in an average of one-quarter of the sun's full energy striking the earth continuously. This result in 342 W on average striking every square foot of surface of the earth.
2. The top of the atmosphere, clouds, and earth's surface reflect slightly more than 30 percent of the incoming light. The reflected amount may increase if

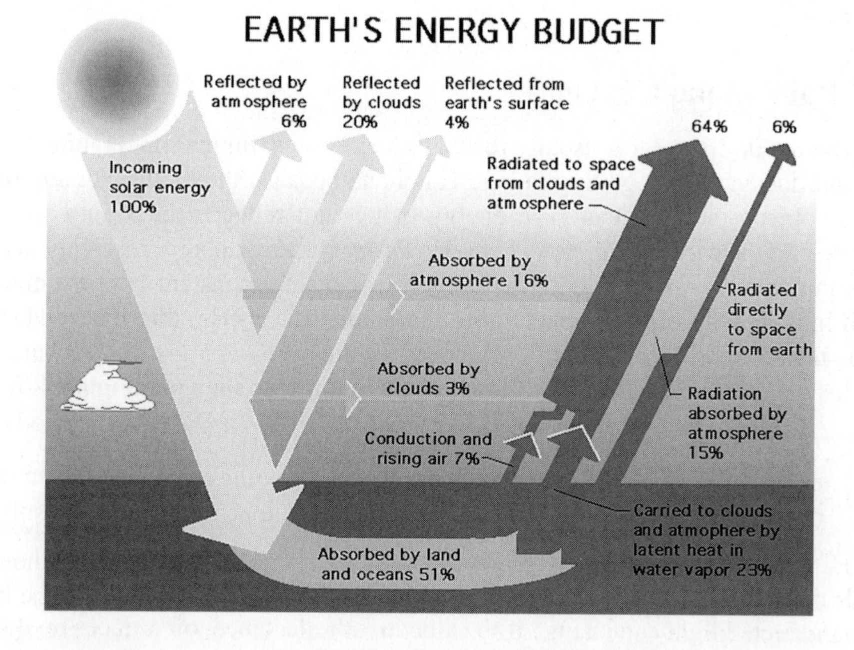

Figure 4-6 Energy paths in the earth's greenhouse. (*Source: NASA.*)

particulates are in the air as a result of volcanic activity. Human-generated pollution also can enhance the amount of reflected light. The remaining 70 percent or so of the incoming light passes through the atmosphere.

3. In the stratosphere, ozone absorbs some of the (short wavelength) ultraviolet light. Water vapor and other gases in the troposphere also absorb some selected colors (wavelengths) from the incoming visible light.
4. Whatever light energy that remains at this point is absorbed by the earth's surface. It heats up land and, in turn, the atmosphere. It drives photosynthesis in plants. It warms the oceans. It melts ice and evaporates water.
5. If the temperature of the earth were absolute zero, it would not radiate heat. However, the temperature of the earth actually is a little above 14°C (58°F). This is not hot enough for the earth to glow visibly, but it does give the earth an invisible thermal presence that radiates infrared light back through the atmosphere. The atmosphere, which is at around –19°C (2°F), also radiates infrared light.
6. Some of the infrared light that the earth radiates is absorbed in the troposphere. Carbon dioxide absorbs selected wavelengths. Water vapor, nitrous oxide, and other gases absorb their particular wavelengths. The infrared energy absorbed in the greenhouse gases of the troposphere causes it to be at a higher temperature.

Maintaining a Balance

On a sunny day, the earth usually reaches a certain maximum temperature for the day but does not keep getting hotter. This is so because the earth, like any other heated object, radiates heat. The energy that is not reflected back into space is absorbed in the earth's climate system. However, if the earth kept receiving heat, it would continue to heat up continuously. To maintain a constant temperature, the earth must radiate back to space the same amount of energy that it receives. By doing this, the earth maintains thermal equilibrium

This balance is based on the following three things, as shown in Figure 4-7.

1. *Incoming solar radiation.* Currently, an average of 342 W is received on every square meter of surface on which the sun shines directly. The amount of incoming radiation can change if the sun's output varies or as a result of changes in the earth's orbit.
2. *Reflection.* Currently, about 30 percent, or 107 W/m^2. The amount of reflected light can change if the albedo, or reflectance, of surfaces on the earth, including clouds, aerosols, ice, or vegetation, varies.

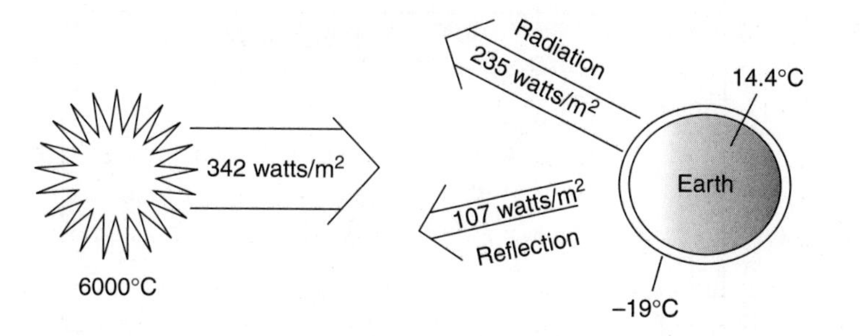

Figure 4-7 Maintaining a balance between light from the sun, reflection, and radiation.

3. Absorption. Currently, about 70 percent, or 235 W/m^2. The amount of invisible infrared radiation absorbed by the atmosphere depends on the concentration of greenhouse gases.

What If the Earth Did Not Have an Atmosphere—Nature's Greenhouse?

Thermal equilibrium is established between the sun, sending about 1370 W for every square meter of surface, and the earth (being a sphere), receiving about 342 W. Thermal engineers know that the temperature of an object floating in space—such as a satellite—will be determined by the heat input from the sun, the amount of light reflected, and the amount of heat emitted. The sun, like any other random object floating in space, is obligated to follow the same laws of thermodynamics. Based on its thermal properties, a 30 percent reflective, 100 percent emissive object in earth's orbital position should be –19°C (–2°F). Without an atmosphere, the earth would be much colder than it actually is.*

This, of course, is *not* the average temperature of the earth. The reason for the difference is that the atmosphere absorbs and retains heat. The reason that the earth is not an uninhabitable slushball is because of the *natural greenhouse effect*.

---

*The amount of energy, E, emitted by a warm object at temperature, T (in kelvins), is given by the Stefan Boltzman's law: $E = \sigma T^4$, where $\sigma = 5.67 \times 10^{-8}$ W/m^2 K^4. At equilibrium, the energy received by the earth equals the energy given off by the sun: $\pi R_E^2(1 - \alpha)S = 4\pi R_E 2\sigma T^4$ where $\pi = 3.14$ and R_E is the radius of the earth (which cancels out of the equation). With albedo $\alpha = 0.3$ and solar constant $S = 1370$ W/m^2, solving for T gives –19°C (–2°F). The earth is much warmer than this calculated value because of the natural greenhouse effect.

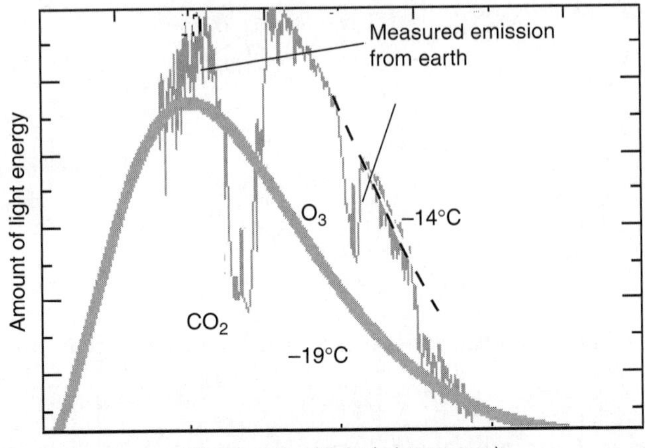

Figure 4-8 Infrared radiation from the earth observed by satellite. Some of this radiation was absorbed by the greenhouse effect.

Satellites Detect Earth's Infrared Radiation from Space

When an object gets warm, we can actually "feel" it from a short distance away. What we feel is the heat radiating across space. (During radiative heat transfer, we would feel the warmth even if there was no air between us and the object.)

Figure 4-8 above shows the glowing earth seen by the eyes of a satellite. The satellite has instruments that work like the infrared glasses that Jack Bauer wears in the television series *24*.

The satellite did not measure that smooth, uniform curve shown in Figure 4-4 which shows the expected temperature range from the earth's surface, 14.4°C (58°F), to the upper atmosphere, −19°C (−2°F). Instead, we see a raggedy line with big chunks missing. Where did that part of earth's heat energy go? The "missing" energy was gobbled up by carbon dioxide and other gases. Ozone also took a bite out of the energy at 10-μm wavelength.

The point of this is that the energy missing from the satellite measurements has been absorbed by the atmosphere and is raising the temperature of the earth. This graph is a direct snapshot of the earth subjected to the greenhouse effect.

Venus and Mercury

Our solar system provides an opportunity to observe how carbon dioxide in a planet's atmosphere can affect its temperature. Mercury is, on average, 58 million km (33 million miles) from the sun, and Venus is, on average, 108 million km (67 million miles) from the sun. Mercury receives about 3½ times more solar radiation than

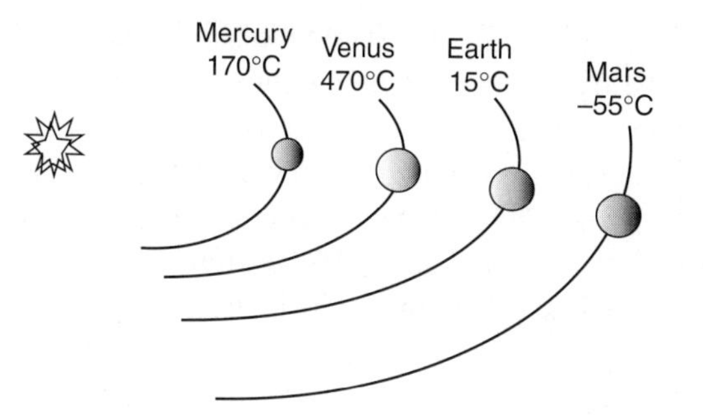

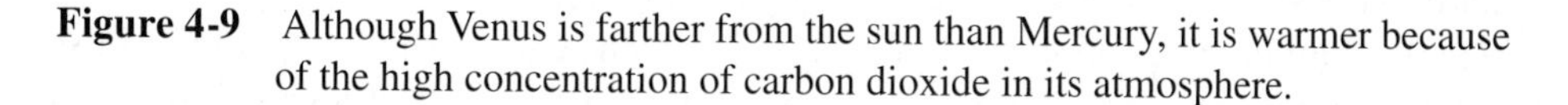
Figure 4-9 Although Venus is farther from the sun than Mercury, it is warmer because of the high concentration of carbon dioxide in its atmosphere.

Venus. Based on distance alone, we would expect mercury to be more than 35 percent hotter than it actually is. It turns out, though, that Venus, despite being farther from the sun, is warmer than Mercury, as shown in Figure 4-9. The average temperature on Venus is over 460°C (860°F), which is far greater than the 170°C (338°F) found on Mercury. The temperature on Venus is hot enough for rocks to glow visibly and for lead to melt. The reason is that the atmosphere of Venus is 95 percent carbon dioxide, whereas Mercury's atmosphere is very thin, with negligible amounts of carbon dioxide. Venus is as hot as it is because of a runaway green house effect taking place there. Venus does not represent a realistic scenario for what might happen to the earth, but it does help us to underscore the influence that absorption in the atmosphere can have on temperature.

Carl Sagan, an American scientist working in the 1960s, determined that the atmosphere of Venus is extremely hot and dense. He related global warming on the earth as a growing, human-induced danger analogous to the transformation of Venus into an inhospitable, overheated planet as a result of the buildup of greenhouse gases in its atmosphere. Sagan's predictions about the surface of Venus were confirmed by the *Mariner 2* spacecraft, whose mission he helped plan.

James Hansen, a NASA scientist, refined Sagan's calculations and included the effects of the sulfate aerosols in Venus's atmosphere. Carbon dioxide is an invisible gas. Sulfates and not carbon dioxide are what gives Venus its characteristic hazy cloud cover. Aerosols such as suspended sulfates play a role in the heat balance in the atmosphere of both Venus and earth.

The Glass of the Earth's Greenhouse

In 1958, Charles Keeling began monitoring carbon dioxide levels in the atmosphere. He chose the remote location of Hawaii's Mauna Loa Observatory to eliminate possible effects of local industry or surrounding vegetation cycles. The observatory was at an elevation 3.35 km (11,000 ft) above sea level, which put it at about the halfway point into the troposphere. Keeling chose that location because he believed that it would give an excellent representation of the entire earth. He developed a method of collecting air samples in flasks and then analyzing them in the laboratory to a level of precision of *parts per million* (ppm). Keeling is credited with having extended the state-of-the-art of measuring small gas concentrations at the time to accomplish this.

Since carbon dioxide is stable in the atmosphere for extended periods of time, and because it mixes very thoroughly with other atmospheric gases, Keeling considered measurements above the Pacific to be truly global in nature. When Keeling started making measurements in 1959, the carbon dioxide level was 316 ppm. This is about 13 percent higher than preindustrial levels. Today, the carbon dioxide reading is over 380 ppm, representing a 35 percent increase above preindustrial levels. The results of these measurements are summarized in Figure 4-10.

Keeling's measurements show a steady increase in the carbon dioxide level. A sawtooth overlay shows a recurrent cyclic increase and decrease each year coinciding

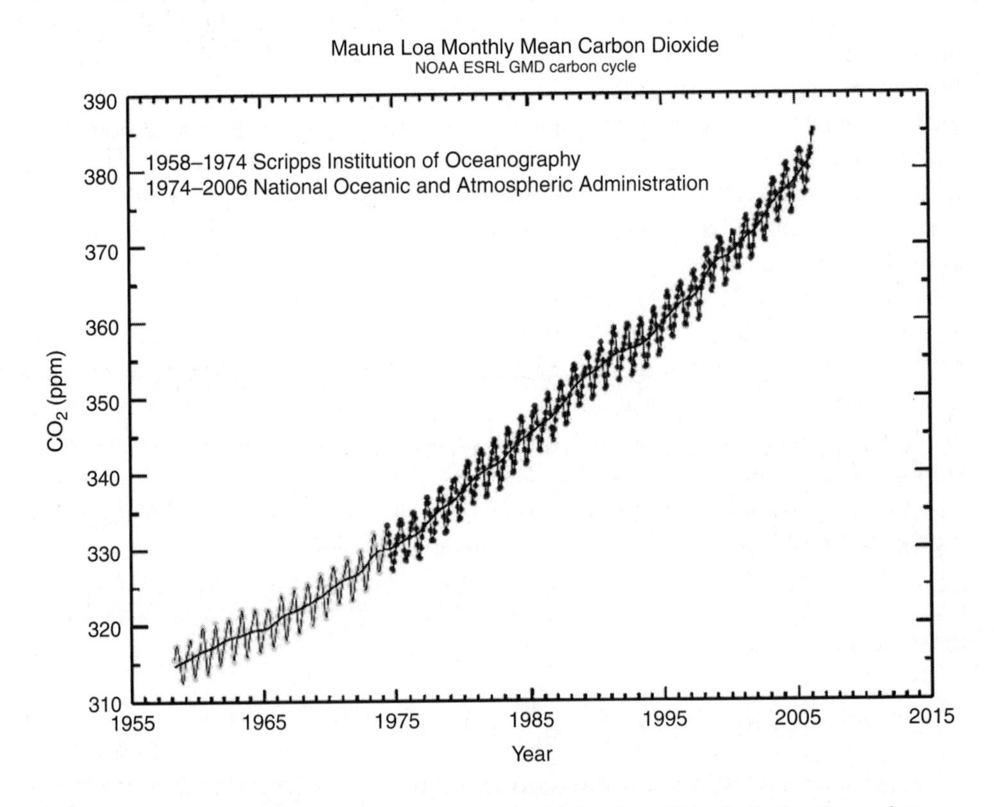

Figure 4-10 Charles Keeling's measurements of carbon dioxide in the atmosphere. (*Source: NOAA.*)

with the alternating growing and winter seasons in the northern hemisphere (which has a much larger presence of plants than the southern hemisphere). Carbon dioxide was higher during the winter and lower in the summer each year. Keeling's chart showed the role of photosynthesis in removing carbon dioxide from the atmosphere during the warmer months when plants were growing. In the fall, leaves and other remnants from the growth process fall to the ground and decompose, releasing carbon dioxide back into the atmosphere. Despite its ups and downs, the most significant feature of the Keeling curve is that the overall trend is upward.

THREE METHODS OF HEAT TRANSFER

Radiation Radiation is unique in that it does not require contact between the objects transferring heat. The sun heats the earth by radiation through space. The earth keeps from overheating by radiating some of the heat that it has absorbed to space. The earth radiates mostly at a temperature that produces invisible infrared light. The greenhouse effect occurs when some of this radiation is intercepted and absorbed by greenhouse gases in the atmosphere.

Convection Sometimes fluids act a conveyer belts to facilitate the flow of heat from one place to another. Prevailing wind patterns transfer heat to restore an imbalance of temperatures on the surface of the earth. Storms perform the same function, although in a more abrupt and often destructive manner. Ocean currents such as the Gulf stream move enormous amounts of heat. This helps to keep regions such as Europe at a much more temperate climate than they otherwise would be.

Conduction This occurs when there is direct contact between materials such as the atmosphere and the oceans. Conduction proceeds until thermal equilibrium is achieved. This happens when the temperature of both adjoining materials is the same. This can take place even though one of the materials contains more heat than the other. This is the case when the air and the oceans reach equilibrium. They are at the same temperature, but a given mass of water has a much greater amount of heat stored in it than a comparable mass of air.

Carbon dioxide does not absorb *visible* light from the sun.

The enhanced greenhouse effect is the result of gases in the atmosphere (such as carbon dioxide) that absorb invisible *infrared radiation* coming *from the warm surface of the earth.*

REFLECTION—SENDING ENERGY RIGHT BACK INTO SPACE

Some of the light coming from the sun is reflected back into space. The only way that it is possible to see the earth from space, as shown in Figure 4-11, is by means of reflection from the earth's atmosphere and surface. Light reflected from the earth does not contribute to heating the earth. Clouds in the atmosphere and ice on the ground enhance reflection. The term *albedo* refers to how reflective various surfaces are.

Reflection from Ice and Snow

At present, ice permanently covers 10 percent of the earth's total land area—nearly all of which is in Antarctica and Greenland. Ice covers another 7 percent of the earth's oceans—including the Arctic Ocean. In midwinter, 49 percent of the land in North America is covered with snow.

As much as 90 percent of the light hitting ice- or snow-covered areas is reflected back into space. In contrast, only about 10 percent of the light hitting the oceans is reflected. Changes in the snow and ice cover can significantly affect the amount of solar heat that is retained in the earth. The impact of this on climate change would be more severe if the sun's rays were less oblique at the poles, where the most

Figure 4-11 The earth is visible from a satellite because about 30 percent of the light striking the earth's surface and atmosphere is reflected back into space. (*Source: NASA.*)

reflective surfaces are located. The extent of reflective areas of ice and snow plays an important role in establishing the thermal balance of the earth.

Reflection from Land

Changes in land can affect how much of the sun's energy is reflected or absorbed in a particular area. Forests tend to be more reflective than clear-cut regions. As a result, loss of forests contributes to warming.

Reflection of light from land may have played a role in Earth's climatic history. A. Wegener proposed that at one time in the Earth's formative years, the land masses of the continents were located in a more centralized continent that he named *Pangaea*. According to his theory, the continents drifted apart. With the continents grouped together, there was more absorbing land mass than reflecting ocean area near the equator, where the suns rays are most direct. The rearrangement of continents as they slowly drifted over millions of years contributed to a cooling trend in the earth climatic history.

Reflection from Clouds

Although clouds both reflect and absorb energy from the sun, their main role is to reduce the amount of sunlight available by reflecting it back into space. Increased cloudiness caused by global warming is to some extent counterbalanced by more atmospheric water vapor in the air. The clouds reflect more sunlight. This partly offsets the warming effect of infrared absorption by water vapor.

If warmer average global temperatures result in an increase in the cloud cover, the overall effect will be to slow the temperature increase. Thus one change in the environment can have an impact on another in an interactive way. In this case, the increase in clouds (caused by higher temperatures) slows the effect of the warming. This is known as *negative feedback*, a topic that will be addressed in Chapter 7.

Contrails

Cloud-like paths left by airplanes are called *contrails*. These are similar to a wake that trails behind a speedboat. Contrails can be thought of as artificial cirrus clouds that—like other clouds—can increase the reflection of incident solar energy. Contrails (Figure 4-12), once formed, become incorporated into the natural cirrus cloud pattern present on a given day. Regions of high air traffic have been found to have more cirrus clouds than regions of low air traffic. The water droplets in these clouds also can play a minor role as a greenhouse gas by retaining radiated infrared light from the earth. The effects are real, but the intricacies of the competing impacts on climate are not well understood by scientists at this time.

Figure 4-12 Contrails are like artificial clouds that reflect incoming sunlight but which also retain heat emitted by the earth at night. (*Source: NASA.*)

Owing to typically cooler nighttime temperatures, contrails form more on overnight flights. Redeye flight contrails form at a time when they contribute only to keeping the earth warm without being offset by reflection of incoming light. When air traffic controllers suspended flights in the United States on September 11, 2001, meteorologists noticed a warming effect that they attributed to the absence of contrails during that time period. However, this observation is based on limited data and possibly was enhanced by the unusually clear weather that occurred at that time.

The Intergovernmental Panel on Climate Change (IPCC) estimates that contrails contribute no more than 1–3 percent of all impacts on the climate generated by humans. Contrails, however, represent only part of aviation's contribution to climate change. Aviation fuel is a fossil fuel that also contributes carbon dioxide emissions. Airplane flight—despite its other many obvious benefits—is probably the least carbon-efficient commercial form of transportation of people and freight. The US Department of Energy estimates that for every gallon of jet fuel that is burned, over 9,500 kilograms (21,000 pounds) of carbon dioxide are emitted.

Aerosols

Aerosols are suspended particles in the air that are contributed by either humans or nature. Aerosols have an overall cooling effect on climate and include particulates resulting from combustion, soot and ash from volcanic activity, nitrates from agriculture, mineral dust, and chemicals suspended in the air such as sulfates. There are many types of aerosols at various locations within the atmosphere, making an understanding of the effects of aerosols complicated. Aerosols are relatively short-lived in the atmosphere and can influence the earth's temperature not only directly by reflecting sunlight but also indirectly by prolonging the time that clouds persist in the atmosphere.

Aerosols have both a direct and an indirect effect on how sunlight is received in the atmosphere. Aerosols scatter or reflect incoming radiation directly. This can cause either a heating or a cooling effect on the atmosphere. Sulfates, organic and solid carbon from fossil fuel and biomass burning, and mineral dust all contribute to a direct aerosol effect.

Indirectly, aerosols influence the reflectivity and duration of clouds. Cloud formation is facilitated by the presence in the atmosphere of particles that serve as nuclei to initiate the condensation of water droplets. Clouds for the most part reduce the amount of incoming sunlight that could contribute to warming the earth's surface.

Aerosols also play a role in modifying the reflectance, or albedo, of surfaces that are critical to maintaining the earth's overall energy balance. Reflectance of ice and snow is reduced to the extent that absorbing particulates settle on frozen surfaces. Absorption in the oceans can be enhanced by the presence of particulates.

Volcanoes

Volcanic eruptions, such as those of El Chichon in 1982 and Mount Pinatubo in 1991, have had major impacts on the earth's temperature by reflecting sunlight from the upper atmosphere. Let's look at the impact of two significant volcanic eruptions: El Chinchon in 1982 and Mount Pinatubo in 1991 (Figure 4-13).

Following the Mount Pinatubo eruption in the Philippines, the clarity of the atmosphere and global temperatures decreased, as shown in Figure 4-14.

The release of aerosol particles caused a measurable reduction of light transmitted into the atmosphere. This can be seen in the sharp increase in optical thickness in the year of the Mount Pinatubo in 1991. *Optical thickness* is a way to characterize the overall loss of clarity in the atmosphere from an eruption. This initial spike following the eruption is followed by a decline as the particles are removed from the stratosphere by natural processes. It took about 4 years for the atmosphere to recover from the Mount Pinatubo eruption.

After a delay of about a year, the earth's temperature dropped by as much as 4°C (7.2°F). In some places, the temperature decrease was even greater. Temperatures recovered by 1995, along with the decrease in particulates in the air.

Figure 4-13 The eruption of Mount Pinatubo poured aerosols into the stratosphere, causing a global dimming effect that lasted over a year. (*Source: NASA.*)

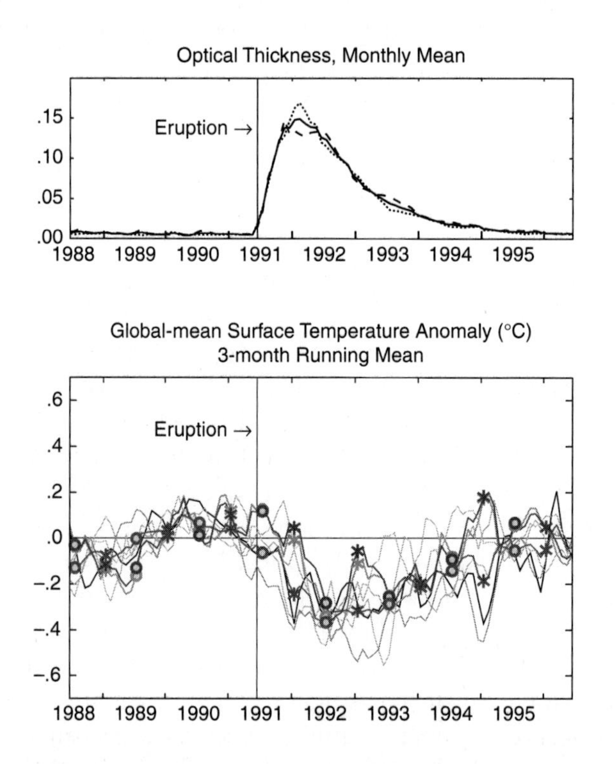

Figure 4-14 Loss of clarity in the atmosphere (optical thickness) and decreased average global temperatures following the Mount Pinatubo eruption in 1991. (*Source: NASA.*)

Every volcano, however, does not have the same impact. For instance, when Mount St. Helens erupted in 1980, its effect on climate was minimal because the plume of ash and other gases and particulates was released primarily into the troposphere, where it was fairly quickly washed out by rain and snow. Only during the first hour did any significant amount of reflective material get disbursed into the stratosphere.

Volcanoes add a large quantity of aerosols into the atmosphere for a concentrated period of time. Their effect on the earth's temperature is significant, temporary, and largely unpredictable.

What about Solar Dimming?

Increased blocking of solar radiation has been referred to as *solar dimming*. During the 1970s, the effects of air pollution were severe enough to cause a small, temporary decline in the overall pattern of global warming. All indications are that global warming would have progressed even more rapidly without the effects of global dimming. The overall trend is toward warming despite the negative influence of increased cloudiness, reflective pollution, and occasional volcanoes.

ABSORPTION—WHERE DOES ALL THAT ENERGY GO?

Where the Earth Keeps Its Energy

Nearly 90 percent of the energy received from the sun goes into the oceans. It may seem odd that with all that heat, the temperature of the oceans has risen only 0.1°C over the past few decades. What happened to all that heat? It takes an enormous amount of heat to change ocean temperature. Not only is the volume of seawater so huge, but water also requires more heat that many other materials on the earth's surface to result in a temperature change. Less than 3 percent of the incoming energy goes into melting ice in glaciers, Greenland and Antarctic ice caps, and Arctic sea ice. Figure 4-15 presents a breakdown of where the sun's energy winds up once it gets to the earth.

Heat and Temperature

Heat and temperature are not the same. Heat is the amount of energy that is contained in a material. Temperature is a measure of how fast the molecules of that material are moving. For example, a cup of tea has a higher temperature but contains less heat than a swimming pool filled with cooler water. Heat

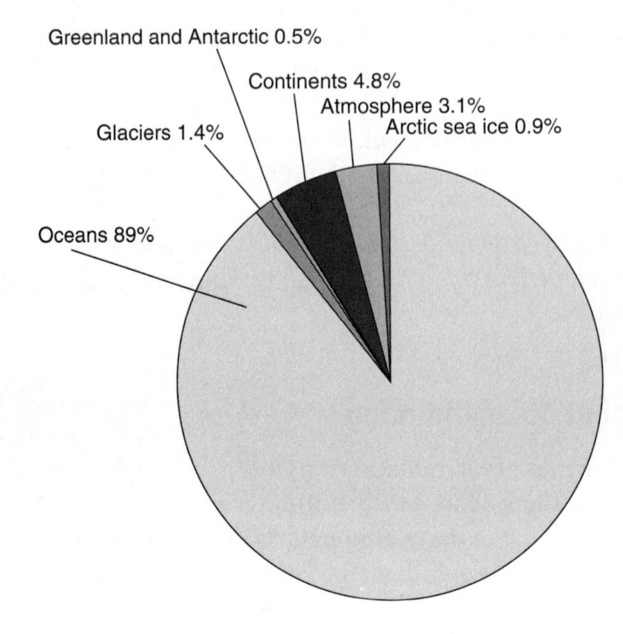

Figure 4-15 Where the sun's energy goes.
(*Source: Based on data from the IPCC.*)

energy is stored in matter by the motion of the molecules of the material. The faster the molecules move, the higher is the temperature of the material. The overall effect of all the molecules moving is the heat energy contained in the material.

Heat Capacity

More heat energy is needed to change the temperature of a given mass of water than to change the temperature of either rock or air. Light energy absorbed by water will change the temperature of that water by a smaller number of degrees than it will change the temperature of air. With close to 70 percent of the earth's surface covered by oceans and lakes, water has a moderating influence on temperature. The *specific heat* of a material is the amount of heat (in *joules*) that it takes to raise 1 kilogram of that material by 1°C. Table 4-1 compares the different effects in three main regions of the earth.

Air gets get hotter four times faster than water (both of the same mass and both exposed to the same amount of heat). The temperature increase for rock when exposed to the same heat energy input is faster than that of either air or water. Without the vast stretches of oceans and lakes covering the earth's surface, the temperature increase from the greenhouse effect would be much more severe.

Table 4-1 How Various Components of the Earth Respond When Heated

Material	Specific Heat, J/kg/°C	Change in Temperature for One Unit of Heat (1000 J) Added to 1 kg of Material
Water	4186	0.22°C
Rock (basalt)	790–840	1.2°C
Air	1006	1.0°C

Latent Heat

Heat energy also can be incorporated into matter in a way that does not change its temperature. This form of heat transfer is called *latent heat*. In the case of ice, it is the amount of heat needed to melt a given mass of material. When ice or any other material melts, it absorbs heat without changing temperature. This means that to the extent that ice is melting, there is no net increase in temperature even though the earth is absorbing more heat. Like water, ice caps and glaciers have a moderating effect on how fast the earth's temperature increases.

The significance of this is as follows:

- If absorption of solar radiation by the earth results in melting ice, the full impact on the earth's average global temperature will be delayed until the ice melts.
- As the ice melts, abrupt changes in local temperature and their associated impact on climate may occur. While an ice sheet is melting, the immediately surrounding ice, water, and air maintain an equilibrium that is close to the melting point of ice, which is 0°C (32°F). Once the ice melts, no more of the incoming heat energy is tied up in the latent heat of melting ice. At that point, any additional heat energy goes into heating the water and air.

Ultraviolet Absorption

Some light coming from the sun is invisible. Light coming into the atmosphere whose waves are spaced apart further than red light is called *infrared light*, which has a longer wavelength than red light. The earth also receives some ultraviolet light, which has an even shorter wavelength than blue light. Light whose waves are closer together than violet light are called *ultraviolet light*. Some, but not all, of this ultraviolet light is absorbed in the stratosphere by the ozone layer. This results in thermal absorption in the stratosphere which is readily reradiated back out to space. Absorption of light radiation in the stratosphere has an overall cooling effect on the surface of the earth since less energy penetrates to that level.

Natural Climate Cycles

SUNSPOT CYCLES

In order to identify how much human activities are influencing the temperature of the earth, it is important to isolate changes that would be taking place without the presence of humans. Accurate measurements of solar output date back only as far as 1978, when satellites began measuring solar intensity outside the interfering influence of the earth's atmosphere. Details about how these satellites monitor the various aspects of the earth's climate can be found in Appendix C. Satellite measurements are accurate enough to detect variations associated with the sun's rotation on its axis about every 27 days. This variation is much smaller than the 11-year solar activity cycle associated with maximum and minimum sunspot periods. Solar radiation is slightly lower when there are fewer sunspots.

Satellite data found that the solar "constant" mentioned earlier is actually not perfectly constant. Instead, it fluctuates in an approximate 11-year cycle along with the ebb and flow of sunspots. This causes a minor ripple in the earth's temperature, which rises and falls by as much as 0.1 percent approximately every 11 years. This variation is accounted for in the overall earth's energy budget but is far too insignificant to be responsible for the observed increases in global air temperature. Figure 4-16 shows how the solar intensity cycles are synchronized with sunspot count and solar flare activity cycles.

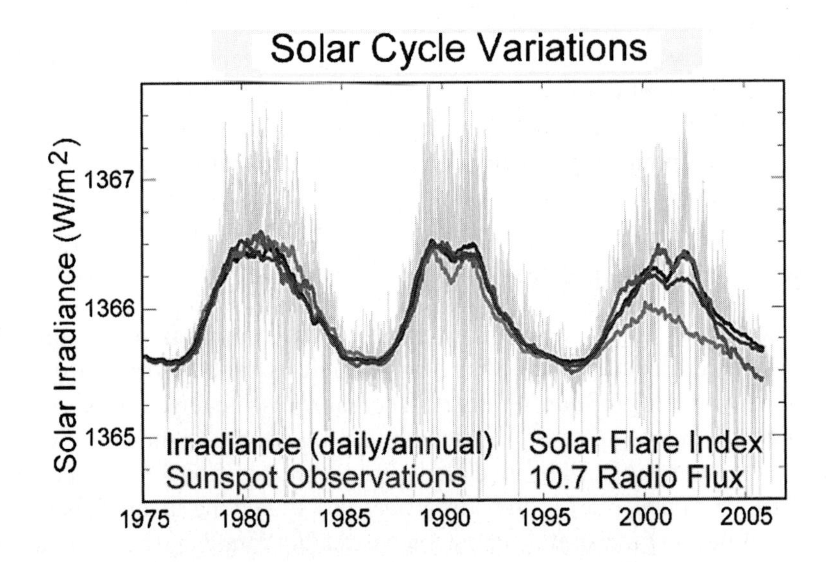

Figure 4-16 Solar intensity fluctuates slightly along with the (approximately) 11-year sunspot cycle. (*Courtesy of R. Rohde, Global Warming Art.*)

During the seventeenth century, an astronomer named Edward Maunder noticed that there were much fewer sunspots at the time than usual. The *Maunder minimum* is the name given to the period roughly from 1645 to 1715 when very few sunspots were recorded and coinciding with a decrease in solar intensity and significantly colder temperatures on earth. Climate records from that time suggest that the sun was 0.15–0.3 percent less bright than the present day. Some scientists attribute this period of colder temperature known today as the "little ice age" to the reduced solar output.

The sun continuously projects a stream of charged particles, called the *solar wind,* at high velocity toward the Earth. When sunspots are at a maximum, the solar wind is most intense. The solar wind affects the electronics and solar panels on satellites and, in the most severe instances, disrupts power and communications on earth. The solar wind is also responsible for the display of shimmering lights called the *northern* and *southern lights* (*aurora boreolis* and *aurora australis*). Some scientists see the variation in intensity of the solar wind as a possible influence on some of the ongoing recurrent climate cycles on earth (K. Frazier, *Our Turbulent Sun,* Prentice Hall, Englewood Cliffs, N.J., 1982).

MILANKOVITCH CYCLES

In Earth's path around the sun, the shape of its orbit and the tilt of its axis of rotation undergo small but significant changes over time. These changes are the result of the gravitational pull of other planets in the solar system, most notably Jupiter. A Serbian mathematician named Milutin Milankovitch (Figure 4-17) completed a study published in 1930 of these changes and their impact on the earth's climate.

Figure 4-17 Milutin Milankovitch.

Milankovitch thought that small slowly evolving changes in the earth's orbit around the sun would eventually lead to an ice age. Based on his orbital calculations, Milankovitch predicted that ice ages should occur every 100,000 years, with smaller temperature swings occurring every 41,000 and 19,000–23,000 years.

Recent evidence from past climate records, including ocean core samples and coral reef measurements, show that Milankovitch's predictions were quite accurate, with ice ages peaking just about every 100,000 years. The three conditions contributing to the Milankovitch cycles are described below.

Eccentricity

Like all the planets in the solar system, the earth follows an elliptical path in its annual orbit around the sun. Every 100,000 years, the earth's orbit goes through a cycle that brings it from a nearly circular orbit to one with a slightly greater eccentricity. The very slight gravitational pull of other planets in the solar system, especially Jupiter and Saturn, on the earth are thought to be responsible for this slight orbital adjustment. The *eccentricity* of an ellipse is a measure of whether the ellipse looks more like a circle or more like an oval. This affects the average global temperature because it determines how far the earth is from the sun during each of the seasons.

Currently, the difference between the closest and furthest distance from the sun is 3.5 percent, translating to a 6.8 percent variation throughout a yearly cycle of solar intensity. (*Note:* Solar intensity varies as the inverse square of the distance between the earth and the sun.) During the most highly elliptical orbit, there would be a variation of 23 percent in the solar intensity during the earth's yearly path around the sun. Figure 4-18 illustrates the changes in the earth's orbital eccentricity.

Axis of Rotation Angle

The earth rotates on an axis that is tilted at an angle of 23.4 degrees to the plane of the earth's orbit. This angle—which is also known as *obliquity*—can vary slightly over time through a range of 2.4 degrees. The earth's axis shifts between a tilt angle of 22.1–24.5 degrees and back again over a period of approximately 41,000 years, as shown in Figure 4-19.

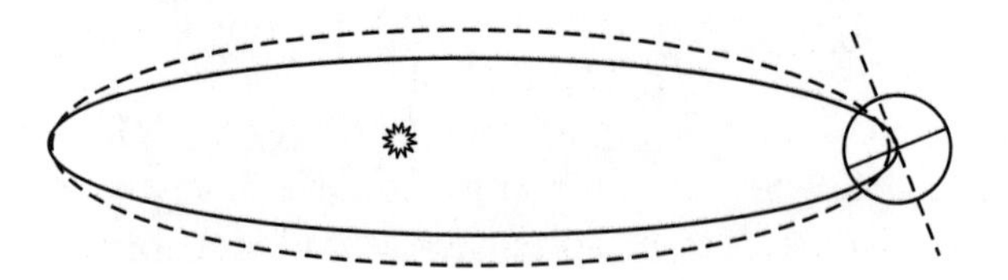

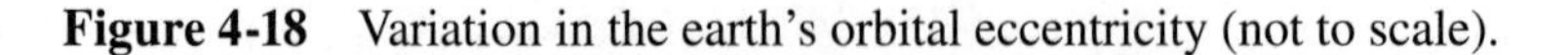

Figure 4-18 Variation in the earth's orbital eccentricity (not to scale).

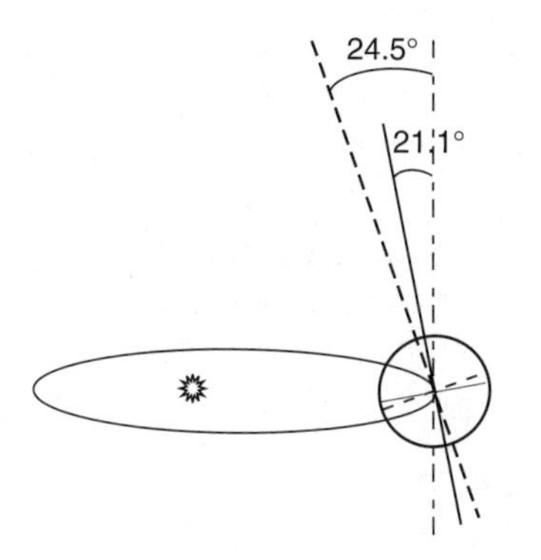

Figure 4-19 Variations in the earth's axis of rotation over time.

When this angle is at it greatest, the earth becomes hotter in summer and cooler in winter. A smaller angle would result in a smaller range of temperature extremes between the seasons.

During the parts of this cycle when the tilt angle is smaller, cooler summers occur near the poles. As a result of less melting of the previous winter's ice and snow, cooler summers lead to a continual cumulative buildup of ice. This creates a condition that favors the start of an ice age.

Currently, the earth is tilted at 23.4 degrees from its orbital plane, about midway between the maximum and minimum angles. The tilt is decreasing slowly and will reach its minimum value in about 8000 years.

Wobble—Precession of the Equinoxes

As the earth spins, it wobbles slightly like a top. This wobbling is known to astronomers as *precession* and is the change in the direction of the earth's axis of rotation with respect to the earth's orbital path around the sun. Therefore, not only does the tilt angle get larger and smaller, but the axis also points in different directions. The direction that the earth's axis spins goes through a full circle roughly every 19,000–23,000 years. The driver of this top-like motion is the force exerted by the sun, moon, and planets on the earth, which is not a perfect sphere.

As we saw in Figure 4-1, at present, the earth is closest to the sun during the winter season in the northern hemisphere. The wobbling eventually will bring the earth's axis of rotation to a point where the northern hemisphere is tilted *toward* the sun at a time when it is *closest* to the sun. This will cause the summer in the northern hemisphere to get hotter and the winter to be colder. As with variations in the axis of rotation, one hemisphere will have a greater variation between the seasons, whereas the other hemisphere will have milder seasonal

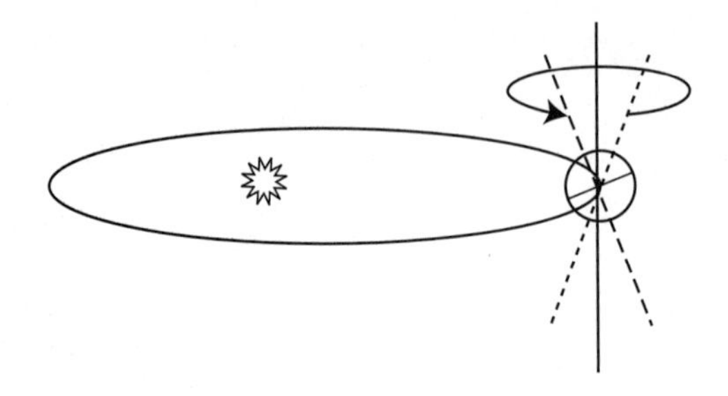

Figure 4-20 Precession of the earth's orbit.

differences. The changes in the earth's rotation associated with precession are shown in Figure 4-20.

The major components of the Milankovich cycles are summarized in Table 4-2.

How This Affects Climate

The dominant effect that triggered the ice ages was orbital conditions that favored reduced sunlight during summer seasons in the northern hemisphere. This resulted in accumulation over time of increased reflective layers of ice that promoted further cooling.

The Milkanovich cycles provide a good explanation of the historical 100,000-year cycle in ice ages. The combined effect of these natural cycles is a variation in the sun's energy that reaches the earth at different times and in different locations. Figure 4-21 shows the Milkanovich cycles plotted in relation to the patterns of glaciation. These cycles result in variations in the overall solar intensity received in the northern hemisphere shown as *solar forcing*. (More about this can be found in Chapter 7.)

These natural climate changes are significant, but they are distinct from the effects of global warming caused by the enhanced greenhouse effect. This natural

Table 4-2 Summary of Milkanovich Cycles

Condition	Range of Variation	Aspect of Climate Most Affected	Interval
Orbit shape (eccentricity)	More circular to more oval shaped	Average distance to sun	100,000 years
Tilt of rotation axis (obliquity)	22.1–24.5 degrees	Severity of seasons	41,000 years
Wobble (precession)	360 degrees	Timing of seasons and proximity to sun	19,000–23,000 years

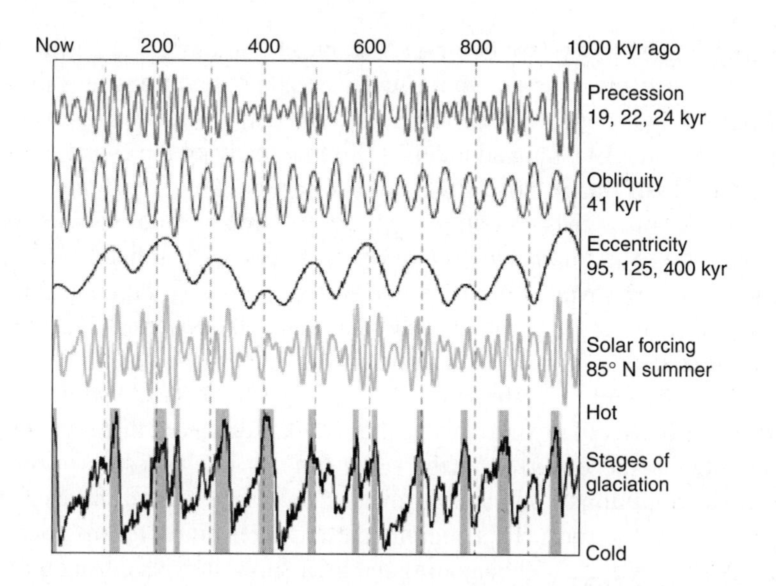

Figure 4-21 Milankovitch cycles compared with recurrence of ice ages (glaciation). (*Courtesy of R. Rohde, Global Warming Art.*)

climate change is exerting a slow *cooling* trend that will take tens of thousands of years to occur. However, any thought that global warming will be offset by the next Milankovitch cycle should be put to rest because the next naturally occurring ice age is not expected for at least another 30,000 years.

> Natural cycles such as those involving the earth's orbits have contributed to initiating past climatic changes. They are not, however, responsible for the global warming that is taking place today.

EL NIÑO AND LA NIÑA/ENSO

Climate Oscillations—Swinging between Two Extremes

El Niño, a part of the climate system, is a naturally occurring cycle in the equatorial Pacific Ocean that repeats roughly every 2–8 years. El Niño is part of a climate cycle called the *southern oscillation*. The cycle varies between two extreme conditions: El Niño and La Niña. *El Niño* is Spanish meaning "the child." South American fishermen gave this weather phenomenon its name at Christmas time, when milder than normal conditions prevailed in western Peru. *La Niña* means "the sister" and is the other part of the cycle, when the opposite, more severe weather conditions prevail. The term *normal year* in this context is sometimes used to denote conditions that are roughly midway between the El Niño and La Niña extreme points of the cycle.

El Niño and La Niña are closely related. They are both part of a long-term climate pattern that meteorologists call an *oscillation*. El Niño and La Niña are part of an oscillation called the *southern oscillation* that was first noticed by the British meteorologist Gilbert Walker in the 1920s. Walker, who was stationed at the time in India, wanted to understand what caused the monsoon season in the southern part of Asia to be severe or mild. Walker noticed that when there was a strong monsoon season in Asia, there were, at the same time, severe droughts in Australia, Indonesia, and parts of Africa. Looking into this further, Walker observed that high air pressure in the eastern Pacific occurred at the same time as low pressure in the western Pacific. These conditions continuously reversed themselves like a child's playground swing going back and forth. This ongoing pattern determines, for instance, whether people in North America will be getting out their snow shovels during the winter or seeing a relentless barrage of hurricanes during the summer.

The southern oscillation, El Niño, and La Niña are all considered to be part of the same condition. For this reason, climatologists refer to it as the *El Niño southern oscillation* (ENSO).

The mechanism that drives the ENSO, as with other climatic oscillations, is a complex interaction between the atmosphere and the ocean. Air pressure differences affect ocean currents both along the surface and vertically, which affect water surface conditions. It is not always easy for the air to push the ocean, so there can be a delayed reaction from the atmospheric influence until the ocean responds. Once the ocean changes, though, it influence air conditions. This continuing dance between the ocean and atmosphere results in a restoring influence that makes the ENSO a recurring phenomenon.

The ENSO Climate Pattern

During the El Niño part of the cycle, certain regions of the Pacific Ocean become warmer than usual. Trade winds and ocean currents change from their typical patterns. This brings abnormal and at times extreme weather to various parts of the world. During La Niña, the opposite occurs, and the ocean is colder than usual.

Near the equator, the trade winds normally blow from East to West. El Niño conditions weaken or reverse this pattern, to produce a West-to-East flow. El Niño can have a significant impact on the earth's climate. During El Niño events, the average global temperatures can increase by 0.1–0.2°C (0.2–0.4°F). This is different from the global warming caused by greenhouse gases because the ENSO temperature change is part of a repeating cycle rather than a progressively increasing trend.

The ENSO affects climate throughout the world. During El Niño years, winters in North America are milder, the southeastern United States and Peru get more precipitation, and there are fewer Atlantic hurricanes. There is drought in the western Pacific and Australia, which under the most severe conditions leads to forest and brush fires. Because of the trade wind patterns, the sea level is 0.5 m (1.6 feet) higher in Indonesia than in Ecuador, and sea surface temperatures are 8°C (14.4°F) higher in the western than in the eastern Pacific. Figure 4-22 shows sea level difference during an El Niño event as detected by a Ocean TOPography Experiment (TOPEX) satellite. Cooler ocean temperatures produce upwelling off South America, which brings nutrients to the surface. Warm air near Peru rises, producing heavier local rainfall. An air pressure difference between the Pacific island of Tahiti and Darwin, Australia, is the official indicator of the onset of El Niño conditions.

La Niña is the opposite side of the coin. Sea surface temperatures near the equator are as much as 4°C (7°F) below normal. Parts of Australia and Indonesia have greater precipitation during La Niña years in contrast to the drought that often characterizes the El Niño years. During La Niña, there are drier conditions in the southwestern United States in late summer and winter. The Pacific Northwest is likely to be wetter than normal. La Niña winters are warmer than normal in the Southeast and colder than normal in the Northwest.

The U.S. National Oceanic and Atmospheric Administration (NOAA) tracks ENSO conditions by using a network of buoys along the equator between the 10 degree N and 10 degree S parallels. These buoys measure temperature, water current, and wind. Satellites also measure sea surface temperature and wave height patterns.

The ENSO is a periodic phenomenon that affects climate worldwide. It is not a cause of global warming in the sense that we are referring to global warming in this book. Although the global mean temperature can be higher during El Niño years, the

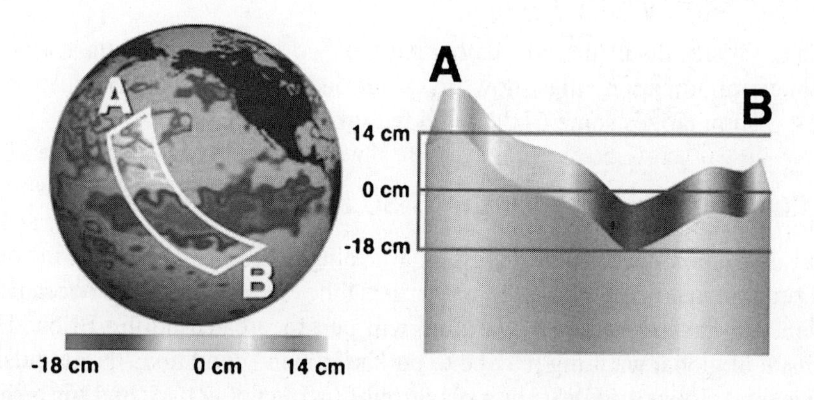

Figure 4-22 El Niño conditions monitored by TOPEX satellite showing differences in sea level. (*Source: NASA.*)

increase is transient. The ENSO is a pattern that—like several others—is superimposed on the overall pattern of global warming. As global warming progresses, El Niño years may be warmer, and La Niña years may be less severe in their effect. El Niño can have devastating effects, producing droughts in some areas and flooding in others. It has completely wiped out the fishing industry in certain areas of the world.

1982–1983 EL NIÑO EVENT

The following are some of the specific effects attributed to this El Niño event:

- An outbreak of encephalitis caused by a warm, wet spring along the East Coast of the United States
- An increase in snake bites in Montana as the hot, dry weather drove mice from higher elevations in search of food and water, and the rattlesnakes followed
- An increase in shark attacks off the Oregon coast owing to unseasonably warm ocean temperatures
- An increase in cholera in Bangladesh
- An increase in typhoid, shigellosis, and hepatitis in South America
- An increase in viral encephalitis in Australia

(*Source: NASA.*)

The ENSO results in altered weather patterns worldwide producing a wide range of climate modifications that show up in various regions during specific seasons. Table 4-3 summarizes some of the main features of the ENSO.

Relationship between the ENSO and Climate Change

Scientists are looking into whether global warming is helping to promote the onset of ENSO conditions. There is also the concern that the effects of increased ocean surface temperatures caused by global warming will be exacerbated during El Niño years. The result of global warming may be to put the El Niño conditions on steroids.

If we think about the ENSO as a playground swing going back and forth between the conditions of El Niño and La Niña, global warming is like someone pushing that swing. Global warming is pushing the climate to the warmer side of the ENSO.

Table 4-3 Characteristics of ENSO Climate Conditions between the Two Extremes of El Niño and La Niña

Condition	El Niño	La Niña
Sea surface temperatures in the eastern Pacific	0.1–0.2°C (0.2–0.4°F) higher	Lower sea surface temperatures
Trade winds	*From the west:* Reversal or slowing of typical trade winds	Typical westerly winds (*from the East*)
Hurricanes	Fewer	More frequent—the 1998 La Niña season was the deadliest in two centuries
Precipitation	Increased rainfall throughout the Pacific coast of northern Peru and Ecuador with summer flooding and drought in parts of Australia and Southeast Asia	Drier in the Southwest and central plains in summer
Winters—North American	Milder winters in the Midwest; California, northwest Mexico, and the Southwest are wetter and cooler than normal	Colder winters in the Northwest; warmer winters in the Southeast
Oceans	Nutrient-rich cold water upwelling to the surface, enhancing fish populations	Normal ocean layering
African climate	Wetter in the East and drier in Central and southern areas	More typical seasonal climate
Antarctic	More sea ice	Less sea ice

The oscillation still occurs, but it is becoming more biased toward warmer ocean surface conditions. The results are shown in Figure 4-23.

OTHER CLIMATIC OSCILLATIONS

The ENSO is just one of several similar recurring climate patterns. These include the *North Atlantic oscillation* (NAO), the *northern annular mode* (NAM), the *southern annular mode* (SAM), the *Pacific–North American pattern* (PNA), and the *Pacific decadal oscillation* (PDO). Like the ENSO, these patterns come and go and interact in ways that are actively being studied by climatologists around the world. Natural climate cycles often last 10–20 years. Changes in climate must be tracked for at least several decades so as not to be confused with the effects of these natural cycles. Interpretation of possible relationships between global warming and local weather events needs to be done in the context of naturally occurring climatic cycles.

For instance, the coming and going of severe weather is often brought on by these oscillations. Many of the graphs in this book show an up-and-down

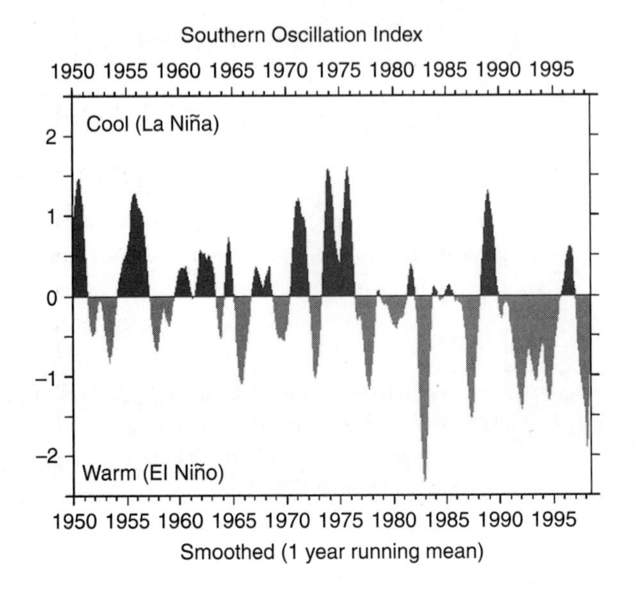

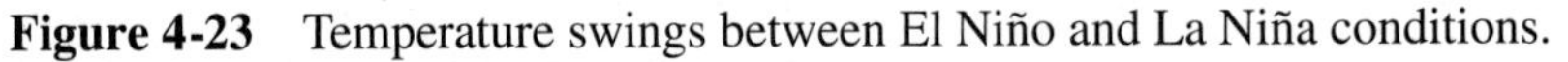

Figure 4-23 Temperature swings between El Niño and La Niña conditions.

roller-coaster pattern superimposed on a long-term trend. The up-and-down pattern is the effect of the oscillations.

> Average global air temperatures take a few steps forward and a few steps back on their way up because of natural climate cycles such as El Niño. We should be careful not to blame a particularly devastating storm season or an unusually warm winter on global warming.

A summary of the major natural climatic oscillations is given in Table 4-4.

Table 4-4 Major Natural Modes That Influence Weather and Climate

Climate Mode	Generally Associated Time Interval
Southern oscillation (SO) (includes the ENSO)	Recurs every 2–8 years Persists 6–18 months
Atlantic oscillation (AO)	Persists weeks to months
Pacific decadal oscillation (PDO)	20- to 30-year cycle
North Atlantic decadal oscillation (NAO)	6- to 10-year cycle
Northern annular mode (NAM)	On the order of weeks
Southern annular mode (SAM)	On the order of weeks

Distribution of Heat around the Earth—Thermohaline Circulation and the Gulf Stream

The *thermohaline circulation* (THC) is a global ocean current that transfers massive amounts of heat from near the equator to higher latitudes in a process often compared with a conveyor belt. Figure 4-24 gives a sense of the paths that this enormous ocean current takes. This circulation is driven by a combination of cold temperatures and high salinity near the ocean surface. The THC is synonymous with the *meridonal overturning circulation* (MOC).

Every day, the Gulf stream carries the heat equivalent of all the coal that is burned everywhere on the earth for a decade. This is why Great Britain has much milder winters than other places at comparable latitudes, such as Canada and Russia. The THC drives the Gulf stream. The thermal satellite image in Figure 4-25 gives an idea of how heat is redistributed from one part of the world to another. As the Gulf stream swings northeastward across the Atlantic Ocean toward Europe, it called the *North Atlantic drift*. Surface waters of the North Atlantic drift approach Europe. There, they grow dense and sink and begin the tedious trek back to the southern hemisphere. This trip takes 1500 years to complete. The massive movement of water and heat has a profound impact on the earth's climate. This circulation also stirs up nutrients from the ocean depths, facilitating ocean life in more places.

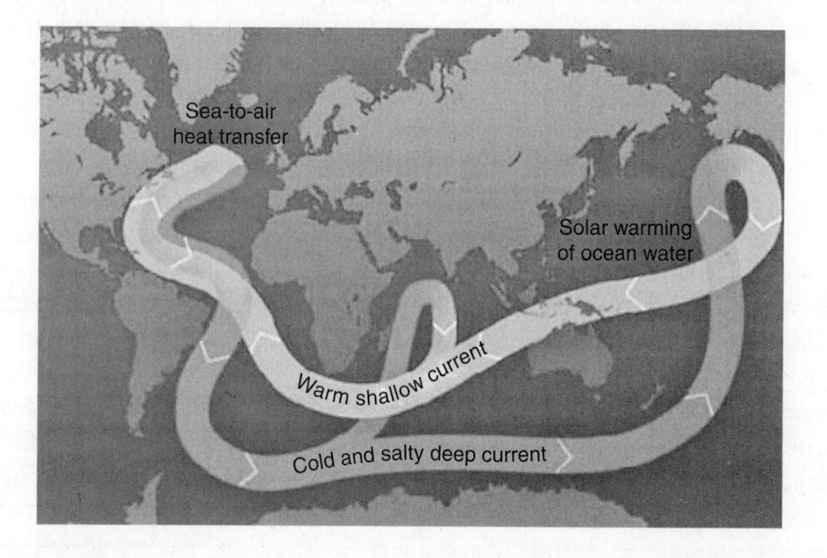

Figure 4-24 The thermohaline circulation. (*Source: NOAA.*)

Figure 4-25 The Gulf stream redistributes heat.
(*Source: NASA Goddard Space Flight Center.*)

Ocean currents (such as the Gulf stream and the North Atlantic drift), air currents (such as the jet stream), and local wind conditions are nature's way of redistributing heat. The energy transported by the Gulf stream is equivalent to the power output of 1.5 million state-of-the art nuclear reactors.

The process of transferring heat by using a moving fluid (such as water or air) is called *convection.* The earth's weather and climate largely are the result of vertical and horizontal convection currents moving in complex paths across its surface. One way scientists know that this is happening is by tracking compounds in the water such as halocarbons, whose presence pins them down to a particular time period. Scientists track how long these trace materials were last in the air and compare their concentration at various depths and locations in the oceans. From this, a map of the THC can be generated.

Ocean circulation patterns are the result of a delicate balance established over thousands of years by the complex interaction of a large number of variables, including salinity and temperature gradients. These circulation patterns may be very sensitive to small changes. Dilution of the ocean's salt content from the accelerated melting of Greenland's glaciers could be the catalyst to initiate this change. Scientists are struggling to see how close we are to the point of no return, but most scientists agree that climate models will need to be much more accurate before we can know for sure. It is not clear how delicate the balance is that keeps the Gulf stream flowing the way is does. Scientists are also asking how much of a change in ocean temperatures would alter the North Atlantic drift current. This current keeps the climate of western Europe much more temperate than that of Labrador, which is at comparable latitude. While the possibility of disrupting this long-established ocean pattern may be real, the likelihood is consistent with only the most pessimistic climate change scenarios.

Key Points

- The sun provides a relatively constant 1368 W for every square meter of the earth's surface.
- This solar "constant" varies by 0.2 percent (±0.1 percent) as part of a natural 11-year sunspot cycle.
- Light is received from the sun in the form of electromagnetic waves. The solar spectrum includes visible, ultraviolet, and infrared light.
- Different colors have different wavelengths. Ultraviolet light has a wavelength that is too short to see. Infrared light has a wavelength that is too long to see.
- The earth's temperature varies throughout the year because of the shape of its orbit around the sun and the angle of its axis of rotation.
- Yearly averages taken over the entire earth eliminate the natural variation that occurs through an annual cycle and at various locations around the planet.
- About 30 percent of the incoming energy from the sun is reflected back into space from clouds, snow, and to a lesser extent other surfaces.
- The balance of the incoming energy from the sun is absorbed by land, air, water, or ice.
- As the earth warms, it radiates invisible infrared light that is partly absorbed by greenhouse gases in the atmosphere, making it warmer. This is called the *greenhouse effect.*
- The greenhouse effect occurs naturally. Human activities have produced an enhanced greenhouse effect as a result of increasing greenhouse gases above historical levels.
- Without the greenhouse effect, the earth would be at the much colder temperature of –19°C (–2°F) instead of the 14.4°C (58°F) it currently enjoys.

- The planet Venus, with a much higher concentration of carbon dioxide in its atmosphere, is a good example of extreme conditions resulting in a greenhouse effect.
- Natural cycles in the earth's motion caused climate changes throughout the earth's geologic history. These changes occur at intervals of 100,000 years in how elliptical the orbit is, 40,000 years in how the angle of the axis of rotation varies, and 11,000 years in variation of the orientation of the axis toward the sun between the seasons. The cycles are known as the *Milanovich cycles* and are responsible for past ice ages and interglacial warming periods.

Review Questions

1. During which (northern hemisphere) season is the earth closest to the sun?
   (a) Spring
   (b) Summer
   (c) Fall
   (d) Winter
2. How is heat transferred from the sun to the earth?
   (a) Conduction
   (b) Convection
   (c) Radiation
   (d) Absorption
3. Compared with blue light, red light has
   (a) a longer wavelength.
   (b) a shorter wavelength.
   (c) the same wavelength.
   (d) a higher frequency.
4. What causes the atmosphere to heat up in the enhanced greenhouse effect?
   (a) Visible light absorbed by carbon dioxide
   (b) Infrared light emitted by the earth and absorbed in the atmosphere
   (c) Visible light reflecting off the surface of the earth
   (d) Infrared light reflecting off the clouds

5. What was S. Arhenius successful in explaining concerning climate change?
   (a) How much the atmosphere heated up for a given increase in carbon dioxide
   (b) The cause of the ice ages
   (c) How much visible light reflected from the troposphere
   (d) The energy driving El Niño
6. What is now the most widely accepted cause of past ice ages?
   (a) Aerosols in the atmosphere from past volcanoes
   (b) A decrease in solar intensity such as during the Maunder minimum
   (c) Increased absorption of carbon dioxide by the oceans
   (d) Milankovitch cycles
7. By about how much did the earth cool off following the global dimming caused by the Tambora and Mount Pinatubo volcanic eruptions?
   (a) 0.3°C (0.5°F)
   (b) 1°C (1.8°F)
   (c) 4°C (7.2°F)
   (d) 8°C (14.4°F)
8. About how much of the radiation from the sun striking the top surface of the earth's atmosphere is reflected?
   (a) 10 percent
   (b) 30 percent
   (c) 50 percent
   (d) 70 percent
9. According to current climate models, how much is the average global temperature expected to increase by, if the carbon dioxide in the atmosphere doubles from present levels?
   (a) 0.5–1.0°C (0.9–1.8°F)
   (b) 2–4.5°C (3.6–8.1°F)
   (c) 8–10°C (14.4–18 °F)
   (d) More than 10°C (18°F)
10. The light emitted from the earth after in absorbs sunlight is in what range?
    (a) Infrared
    (b) Visible

(c) Ultraviolet
(d) Radio waves

11. How often do the ENSO cycles repeat?
    (a) Every 2 weeks
    (b) Every 2 months
    (c) Every 2–8 years
    (d) Every 10,000 years
12. Why did Charles Keeling's carbon dioxide graphs show an up-and-down pattern each year?
    (a) Measurement errors
    (b) More carbon dioxide was generated by furnaces during the winter.
    (c) The air was denser in the winter, causing the carbon dioxide to become diluted.
    (d) Carbon dioxide was absorbed each year during the growing cycle.

CHAPTER 5

Greenhouse Chemistry

The human presence on the earth has developed to the point where we have acquired the ability affect the earth's climate. At one time, the gases our historical ancestors put into the atmosphere by burning wood to cook the animals they hunted was insignificant. The small quantities of greenhouse gases they produced were absorbed and diluted by the air. Today, this is no longer the case. This chapter will introduce you to the greenhouse gases—where they come from, where they go, and how they absorb heat. Some of the components of the atmosphere are natural. Some are natural substances at unnatural concentrations. Some are exclusively a human contribution. Since all fuels—with the single exception of hydrogen gas—have carbon as part of their chemical makeup, every fuel produces carbon dioxide when it burns. (I assume that not all readers "speak chemistry" fluently, so all chemical reactions referred to in this chapter will be described in plain English wherever chemical symbols are shown).

In this chapter we focus on the chemical changes that relate to climate change. This will provide insight into and background for the other chapters in this book. For instance, isotopes are a tool used by climate "forensic teams" to determine the ages of ice-core or tree-ring layers, as discussed in Chapter 2. Isotopes are also used to sort out whether a greenhouse gas is from natural or human origins, which will be explored in Chapter 6. Therefore, while we are on the topic of atoms, this chapter also will focus on what isotopes are and how they are used in climate research.

Gases in the Atmosphere

The following gases make up our atmosphere:

Nitrogen, 78 percent

Oxygen, 21 percent

Water vapor, 0–2 percent

Argon, 0.1 percent

Carbon dioxide, 0.04 percent [280 parts per million (ppm) preindustrial, 379 ppm in 2005]

Methane, 0.0002 percent [715 parts per billion (ppb) pre-industrial, 1774 ppb in 2005]

Nitrogen dioxide, 0.00003 percent (270 ppb preindustrial, 319 ppb in 2005)

Stratospheric ozone, trace amount

Tropospheric ozone, trace amount

Generating Carbon Dioxide

THE CARBON CYCLE

Carbon dioxide enters the atmosphere naturally by

- Anaerobic bacteria that decompose other organic matter
- Animals that exhale carbon dioxide during respiration
- Occasional volcanic activity

Nature removes carbon dioxide from the atmosphere by

- Plants consuming it in the process of photosynthesis
- Water dissolving carbon dioxide

Some of the carbon dioxide that is released into the air is removed by being dissolved in the oceans and other bodies of water. The amount of carbon dioxide that can dissolve in a volume of water depends on the temperature of the water and the concentration of carbon dioxide that is already dissolved in the water. Warmer water dissolves less carbon dioxide than cooler water. Water with a higher concentration of carbon dioxide also dissolves less carbon dioxide than water with little carbon dioxide. This process is self-limiting because as the greenhouse effect causes higher temperatures, the removal rate of the carbon dioxide decreases.

As carbon dioxide dissolves in water, it also causes the water to become more acidic (as discussed in Chapter 3).

BURNING FOSSIL FUELS

The fossil fuels are coal, oil, and natural gas. The name *fossil fuels* derives from the origin of these fuels from the remains of organic matter preserved from prehistoric times. Most of our energy used for electricity and transportation today comes from the burning of fossil fuels, which generates carbon dioxide as a by-product. As a result of fossil fuel combustion, carbon dioxide in the atmosphere is now 35 percent higher than it was a century and a half ago.

Whenever anything containing carbon burns, a product of that combustion is carbon dioxide and water vapor. The basic chemical reaction for the complete combustion of a fuel is

$$\text{Fuel} + \text{oxygen } (O_2) \rightarrow \text{carbon dioxide } (CO_2) + \text{water } (H_2O)$$

In this equation, the arrow stands for *produces or yields*. Sometimes the combustion is not complete, which results in production of carbon monoxide and only partial breakup of the fuel molecules.

Table 5-1 shows the chemical formulas for common fuels. Notice that every one of these fuels except hydrogen contains carbon. If there is a C in the formula, it contains carbon. (H represents a hydrogen atom, and O represents an oxygen atom.) Carbon dioxide has the formula CO_2, which means that it has one carbon atom and two oxygen

Table 5-1 Chemical Formulas of Fuels

Coal (basically carbon with impurities such as water, sand, and sulfur)	C
Natural gas (methane and ethane)	CH_4 and $C2H_6$
Gasoline	C_6H_{14} to $C_{12}H_{26}$
Ethanol	C_2H_5OH
Diesel fuel	$C_{10}H_{22}$ to $C_{15}H_{32}$
Hydrogen gas	H_2

atoms. Oxygen in the atmosphere has the formula O_2. which means that it has two oxygen atoms that make up the gas oxygen that is a constituent of the atmosphere. The atoms you start with get rearranged in any chemical reaction to form new substances with properties that are different from those of the substances you started out with.

COMBUSTION REACTIONS

Combustion of carbon (the main chemical component of coal):

$$C + O_2 \rightarrow CO_2 + H_2O + \text{heat}$$

Combustion of methane (a primary constituent of natural gas):

$$CH_4 + 2O_2 \rightarrow CO_2 + 2H_2O + \text{heat}$$

Combustion of octane (a representative constituent of gasoline):

$$2C_8H_{18} + 17O_2 \rightarrow 16CO_2 + 18H_2O + \text{heat}$$

Combustion of ethanol:

$$C_2H_5OH + 3O_2 \rightarrow 2CO_2 + 3H_2O + \text{heat}$$

Combustion of hydrogen:

$$2H2 + O_2 \rightarrow + 2H_2O + \text{heat}$$

Combustion of diesel fuel (using a typical formula representative of either biodiesel or petrodiesel fuel):

$$C_{12}H_{24} + 18O_2 \rightarrow 12CO_2 + 12H_2O + \text{heat}$$

All fuels (except hydrogen) have carbon in their chemical formula and produce carbon dioxide when they burn.

CEMENT PRODUCTION—PAVING PARADISE

Humans put carbon dioxide into the atmosphere in other ways besides burning fossil fuels. One example is the production of concrete, which involves the conversion of calcium carbonate rock (limestone) into calcium oxide. This process also produces carbon dioxide, as shown in this reaction:

Calcium carbonate + oxygen + heat → calcium oxide + carbon dioxide

Development of cities and towns and industries contributes to global warming in at least three ways:

1. It releases carbon dioxide directly to the air through the production of cement.
2. It requires heat, which is likely to involve the combustion of fossil fuels.
3. It often replaces vegetation that absorbs carbon dioxide from the atmosphere.

The chemical reaction for cement production is shown below.

CEMENT MANUFACTURING

Carbon dioxide is released when limestone is heated to produce quicklime during the production of cement.

$$CaCO_3 + O_2 + \text{heat} \rightarrow CaO + CO_2$$

Absorption of Light by Gases in the Atmosphere

Fossil fuels are a concern because they increase the absorption of incoming solar radiation above the amount absorbed naturally. This is referred to as the *enhanced greenhouse effect.*

When a gas molecule absorbs energy, it has no place to store that energy other than in some form of movement of the molecule itself. According to the *kinetic theory of matter*, molecules are constantly in motion. The molecules move faster when their temperature goes up. Gas molecules in the atmosphere can absorb energy passing through the atmosphere in the form of visible light (from the sun) or invisible electromagnetic waves (from the earth). Some molecules are able to move in more ways than others, enabling them to be better absorbers of certain kinds of light (or electromagnetic) energy.

If we look at the chemical structure of some of the gases in the atmosphere, we can understand better how they absorb light energy passing through the atmosphere. Let's start with nitrogen (Figure 5-1). Nitrogen gas molecules each have two atoms of the element nitrogen bonded together to form a molecule of nitrogen gas that has the formula N_2 (which means two nitrogen atoms).

Because the nitrogen gas molecule is so simple, it cannot do very much with the light energy that it absorbs. It can spin or vibrate only a little bit by stretching and pulling. Oxygen acts pretty much the same way (Figure 5-2).

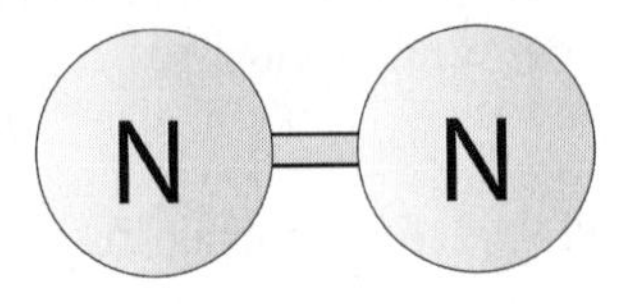

Figure 5-1 Nitrogen molecule.

Because of their structural simplicity, nitrogen and oxygen, which together make up 99 percent of the earth's atmosphere, absorb relatively small amounts of the visible light energy coming from the sun that passes through the air. This leaves the "glass" of the earth's greenhouse pretty clear.

Water vapor, however, is a different story. Water (H_2O) has two hydrogen atoms bonded to a single oxygen atom. The oxygen atoms are bent to form a 105-degree angle (Figure 5-3).

The water molecule can twist, turn, gyrate, bend, flex, and do its own little chemical dance. Because water has more ways to move, it is better able to absorb light energy. Water vapor in the atmosphere absorbs certain colors of light (visible wavelengths) on the way down to the earth's surface as if there were a film on the glass of earth's greenhouse causing it to be less than perfectly transparent. Water also absorbs some of the heat energy in the form of infrared light (long invisible wavelengths) emitted by the earth. For this reason, water is a natural greenhouse gas and contributes to the absorption of heat in the atmosphere. However, as a consequence of global warming, there is more water vapor in the atmosphere, and this, in turn, absorbs even more heat.

The carbon dioxide molecule, like the water molecule, has three atoms. These atoms, rather than being bent, are arranged in a straight line (Figure 5-4). Because of its molecular structure, carbon dioxide happens to be especially well suited to absorb certain wavelengths (mostly the invisible infrared radiation emitted by the earth). It does not strongly absorb the light energy coming from the sun.

Carbon dioxide lets sunlight through the atmosphere but does not let the energy emitted by the earth pass back into space. In this way, carbon dioxide acts a greenhouse gas.

Methane has four hydrogen atoms bonded to a central carbon atom. Looking at the structure of the methane molecule (Figure 5-5), one might (correctly) guess that it has a number of ways to vibrate as it absorbs light (and other invisible electromagnetic)

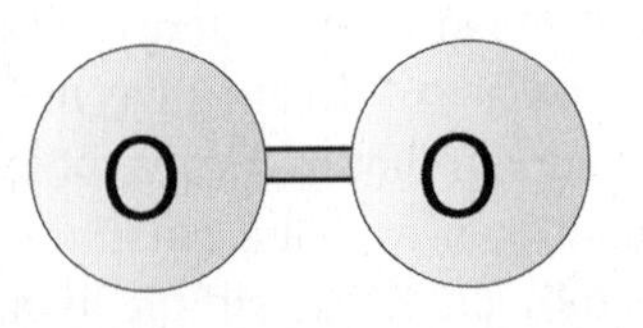

Figure 5-2 Oxygen molecule.

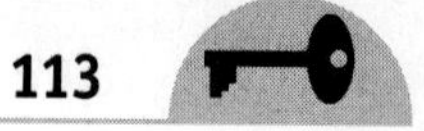

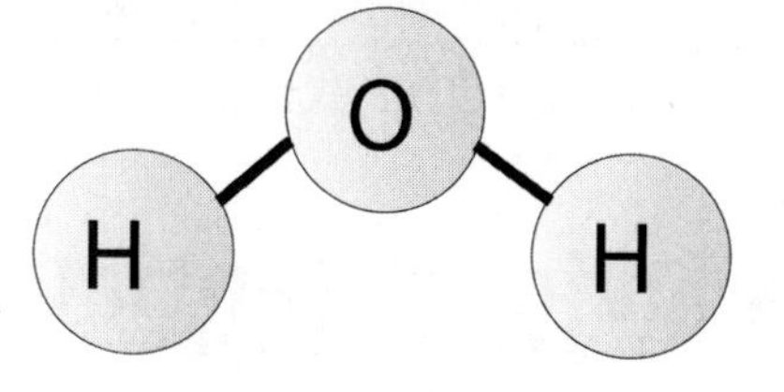

Figure 5-3 Water molecule.

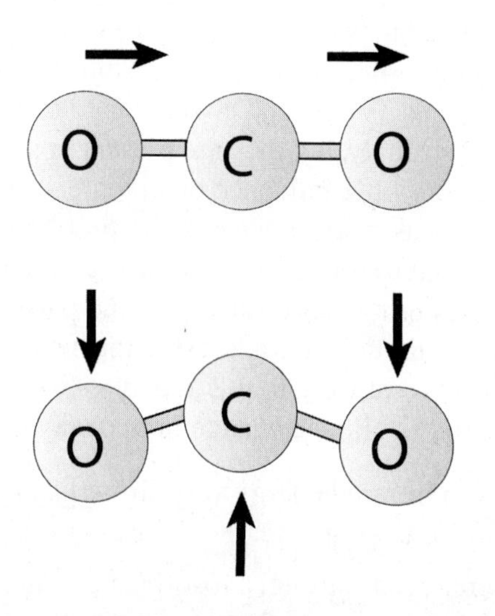

Figure 5-4 Carbon dioxide molecule with the two main vibrational modes that absorb infrared energy.

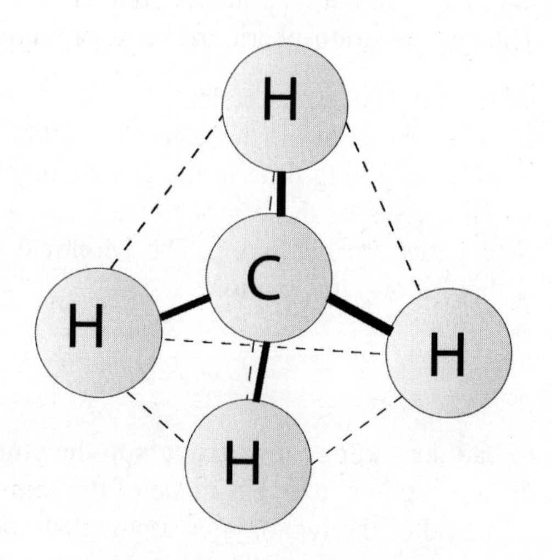

Figure 5-5 Methane molecule.

waves. Like a good dancer, methane has "lots of moves." For this reason, methane is an "energy sponge" that absorbs sunlight 20 times more strongly than carbon dioxide.

Trace Gases with a Major Potential Impact

MONTREAL PROTOCOL GASES

Most of the effects of global warming come from the major human-generated greenhouse gases—carbon dioxide, methane, and nitrous oxide. In addition, there are chlorinated hydrocarbons, ozone, and a wide range of trace molecules. Most of these absorb energy very strongly, but they are present in such small concentrations that they currently have very little influence on climate.

In 1989, a conference was held to address the problem of stratospheric ozone depletion. A total of 191 countries approved the treaty, called the *Montreal Protocol,* that resulted from this gathering. The concern of the participants was not directly global warming at the time. Participants addressed the materials used in refrigeration and aerosol can propellants that were found to deplete ozone in the stratosphere.

These gases are relevant to climate change in two ways:

1. These gases absorb infrared energy very strongly. Despite their low concentrations in the troposphere, these gases contribute to global warming.
2. In the stratosphere, ozone strongly absorbs incoming ultraviolet light. To the extent that this occurs, a small amount of solar energy never makes it to the earth's surface to contribute to global warming. Stratospheric ozone actually contributes a small but measurable cooling effect that may begin to be more significant as the stratospheric ozone layer recovers.

As a result of the participating countries' actions to reduce halocarbon gases, their emission levels decreased by a factor of 3 in the years since 1980. Since, the Montreal Protocol gases persist for a very long time in the environment, only 1–2 percent is removed each year. The levels of these gases in the atmosphere appear to be stabilizing and in some cases are declining. The Montreal Protocol gases are summarized Table 5-2 that follows this section.

OZONE

Ozone can be good or bad depending on where it is in the atmosphere. No matter where it is found in the atmosphere, ozone is made of the same oxygen atoms that make up oxygen gas. Instead of the two oxygen atoms that bond together to form oxygen gas (O_2), however, three oxygen atoms bond together to form ozone (O_3). There is a naturally occurring layer of ozone in the stratosphere that absorbs harmful

Table 5-2 Montreal Protocol Gases and Others

Gas	Destroys Ozone?	Greenhouse Gas?
Chlorofluorocarbons (CFCs, also called *halogenated hydrocarbons*) A few examples include $CFCl_3$, CF_2Cl_2, $C_2F_3Cl_3$, $C_2F_4Cl_2$, and C_2F_5Cl.	Yes	Yes
Hydrochlorofluorocarbons (HCFCs) Used as a replacement for CFCs. A couple of examples include $CHClF_2$ and $CHCl_2CF_2$.	Yes (but 90 percent less)	Yes
Hydrofluorocarbons (HFCs) A couple of examples include CHF_3 and CH_2F_2.	No	Less than HCFCs
Perfluorinated compounds A few examples include SF_6, NF_3, CF_4, and C_2F_6.	No	Yes

ultraviolet light before it can enter the atmosphere. Chemicals used for spray cans and refrigerants, which contain chlorine and fluorine, tend to decompose ozone in the stratosphere. This effect was most notable over Antarctica, where a hole appeared in the ozone layer. When scientists recognized that this was happening, a ban was placed on the use of these chemicals. Since then, the ozone layer has been recovering. This resulted in a slight atmospheric cooling as additional ultraviolet light again began to be absorbed in the upper atmosphere, leaving less to be absorbed closer to the earth's surface. This is the "good" ozone.

The "bad" ozone is the product of combustion and a complex chemical reaction that occurs in the lower layer of the atmosphere, or the troposphere. Ozone in the troposphere is a greenhouse gas that absorbs a small amount of infrared radiation. Compared with the other heavy hitters in the atmosphere, ozone, despite its notoriety of the stratospheric ozone hole, plays only a small role in changing the global temperature.

OZONE

Creating Ozone

Ozone (O_3) is just three oxygen atoms bonded together instead of two, as is found in oxygen (O_2) gas. Natural processes such as lightning convert oxygen gas to ozone. Ozone is also found in tailpipe and smokestack emissions resulting from a more complex photochemical reaction.

$$3O_2 + \text{heat} \rightarrow 2O_3$$

Three molecules of oxygen gas (O_2) produce two molecules of ozone (O_3).

continued . . .

Destroying Ozone

In the reverse of the preceding chemical reaction, the three atoms of the ozone molecule are rearranged to form oxygen molecules. This reaction is catalyzed (or chemically facilitated) by a class of chemicals that contain carbon and chlorine, fluorine, or bromine and a few others that are listed in the next section. There are several more complicated steps in this process, but the basic idea is

$$2O_3 \rightarrow 3O_2$$

Carbon Sinks

DISSOLVING CARBON DIOXIDE

About half the carbon dioxide that humans add to the atmosphere is removed by nature in what is called a *sink*. Had this not occurred, the carbon dioxide levels and the enhanced greenhouse effect would be even greater than it is currently. Nature removes carbon dioxide from the atmosphere by dissolving it in seawater. Dissolved carbon dioxide forms a mild acid called *carbonic acid* that slightly decreases the pH level of the water in which it is dissolved. The amount of carbon dioxide that can be absorbed by the oceans reaches a saturation level depending on water temperature. Warm water can hold less carbon dioxide than cold water.

There is a limit as to how much carbon dioxide can dissolve, and since there is not very much vertical mixing in the oceans, it can take centuries for the dissolved carbon dioxide to penetrate below the surface layers. Although the oceans have an enormous capacity to absorb much of the carbon dioxide that is being added to them, only a small part of the oceans near the surface is available to accomplish this. Consequently, injection of carbon dioxide deep into the oceans has been proposed as a possible way to remove carbon dioxide from the atmosphere.

SEQUESTRATION

One idea for getting rid of some of the excess carbon dioxide is to capture and store it somewhere, such as in the oceans. This is called *sequestration* and will be explored in more detail in Chapter 8 as an option for reducing emissions from power plants. If carbon dioxide from a power plant is pumped into the ocean and released near the sea floor, the carbon dioxide could react with the solid calcium carbonate ($CaCO_3$) found in sea shells to form soluble calcium bicarbonate [$Ca(HCO_3)_2$]. This would

provide long-term storage for carbon dioxide that otherwise would have been released to the atmosphere to contribute to global warming.

REMOVING CARBON DIOXIDE FROM THE ATMOSPHERE

Dissolving Carbon Dioxide in the Oceans

Carbon dioxide dissolves in the oceans and forms carbonic acid. This is a natural part of the carbon cycle:

$$CO_2 + H_2O \rightarrow H_2CO_3$$

Long-Term Storage of Carbon Dioxide Near the Ocean Floor

Carbon dioxide reacts with calcium carbonate (found in sea shells) to form soluble calcium bicarbonate:

$$CO_2 + CaCO_3 + H_2O \rightarrow Ca(HCO_3)_2$$

Integrated Gasification Combined Cycle (IGCC)

IGCC makes it easier to capture carbon dioxide from the smoke stacks of electrical generating plants. First, the coal (C) is reacted with steam (H_2O) to form hydrogen gas (H_2) and carbon monoxide (CO).

$$C + H_2O \rightarrow H_2 + CO$$

Next, the carbon monoxide is converted to carbon dioxide, and the hydrogen is burned to generate additional heat. The carbon dioxide does not get diluted with nitrogen as it is in conventional coal plants and can be removed more easily.

$$CO + 2H_2O \rightarrow CO_2 + 2H_2$$

Comparison of Fuels

We can compare the effectiveness of fuels in several ways:

1. How much carbon dioxide is produced for a given amount of *fuel mass* burned? This is shown in Figure 5-6 for several fuels. For a given mass of

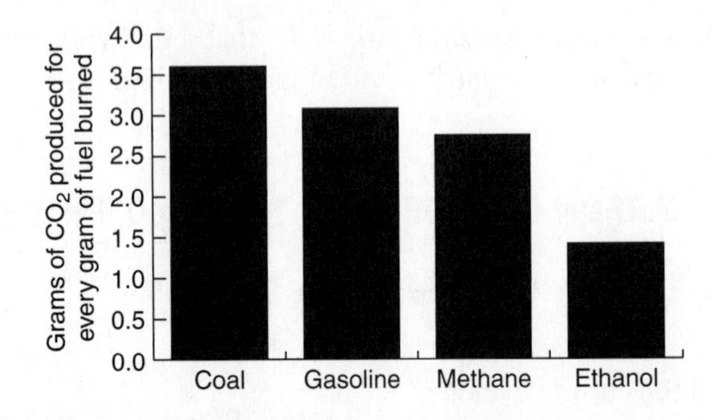

Figure 5-6 Carbon dioxide emissions for a given mass of fuel burned (kilograms of carbon dioxide per kilograms of fuel).

fuel burned, coal produces the greatest amount of carbon dioxide. Methane produces slightly less than gasoline (represented as octane), and ethanol produces the least.

2. How much carbon dioxide is produced for a given amount of *energy* released? This may be a better metric to characterize how "green" a fuel is. It compares how much carbon dioxide is released to generate a given amount of heat energy. These results are shown in Figure 5-7. Coal, which powers the bulk of the world's electricity-generating plants, produces the most carbon dioxide for a given amount of energy released during

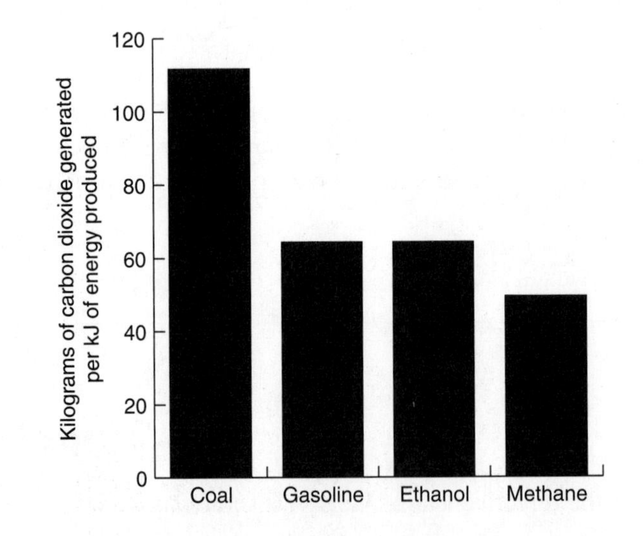

Figure 5-7 Carbon dioxide emissions for a given amount of energy derived from the fuel being burned.

combustion. Replacing coal with any other fuel on the list would cut carbon dioxide emissions for a given amount of heat produced. Ethanol, which is a candidate to replace some of the gasoline used in cars, produces just about the same amount of carbon dioxide as gasoline for a comparable amount of energy produced. This means that ethanol has no inherent benefit as a fuel other than the carbon dioxide that it might pull out of the air during its production. This carbon dioxide is returned right back to the atmosphere when the ethanol is burned.

3. How much energy can we get for a given mass of the fuel burned, or *energy density*? This metric is useful for fuels used for transportation. Using fuel energy to move fuel mass builds in extra inefficiency into the transportation system. On the basis of mass, hydrogen has a very high energy density, as shown in Figure 5-8. This makes hydrogen an attractive potential candidate for transportation systems. Hydrogen, unfortunately in gaseous form has a very low volume density making it less convenient than other fuels to power vehicles. Being a gas, hydrogen would need to be highly compressed or otherwise stored (such as through a hydride) in a small enough volume to keep the vehicle's fuel tank a reasonable size. Ethanol is slightly better than gasoline in terms of mass. However, on the basis of volume, gasoline has a higher energy density. This means that gasoline requires a smaller fuel tank to go a given number of miles than mixes that include ethanol. Coal is the lowest in this group, which helps to explain why coal (with the exception of its use to power locomotives and steamboats), is most widely used as a stationary source of energy.

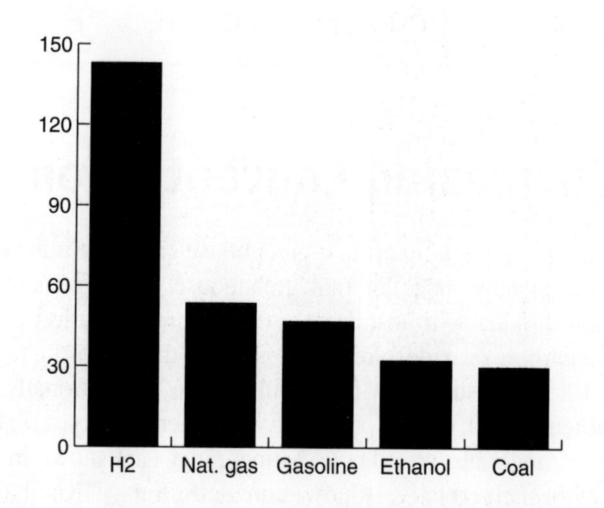

Figure 5-8 Energy density on the basis of mass for various fuels.

It is important to keep in mind that many other factors beside the fundamental heat of combustion evaluated here go into designing a fuel to operate in a real engine. It is equally important to keep in mind the overall life cycle of the fuel. This includes the energy that goes into producing and distributing the fuel and the carbon dioxide released during those processes.

Hydrogen is a unique fuel that does not produce carbon dioxide when it is burned. As indicated in the preceding box showing the combustion products of various fuels, hydrogen generates only water. Finding an effective way to produce hydrogen would give it a more prominent role in the future energy mix. One particularly appealing idea is to use electricity generated by renewable sources to produce hydrogen by electrolysis. The chemical reactions are given in the following box.

PRODUCTION OF HYDROGEN

By electrolysis of water: Hydrogen can be produced by decomposing water using electric current or very high temperatures:

$$2H_2O \rightarrow 2H_2 + O_2$$

This reduces carbon dioxide emissions only to the extent that the electricity used for the electrolysis is not derived from burning fossil fuels.

From methane: Hydrogen can be produced by reacting methane (CH_4) with water (H_2O) in the form of steam. This consumes methane, requires energy, and produces carbon monoxide as a by-product:

$$CH_4 + H_2O \rightarrow CO + 3H_2$$

The Effect of Increasing Concentration

The impact of increasing the amount of a greenhouse gas in the atmosphere on global warming depends on how much of that greenhouse gas is present to begin with. Additional carbon dioxide will absorb only the infrared radiation that has not been absorbed by the carbon dioxide already present in the atmosphere. Adding a few molecules of a trace gas such as $CFCl_3$ will have a proportionally greater impact because those molecules will not be competing with others of its type for the wavelengths that the gas most strongly absorbs. Three distinct ranges are shown in Figure 5-9.

Consistent with their relatively low concentration, CFCs have the highest absorption fraction. Absorption nearly doubles for a doubling of concentration.

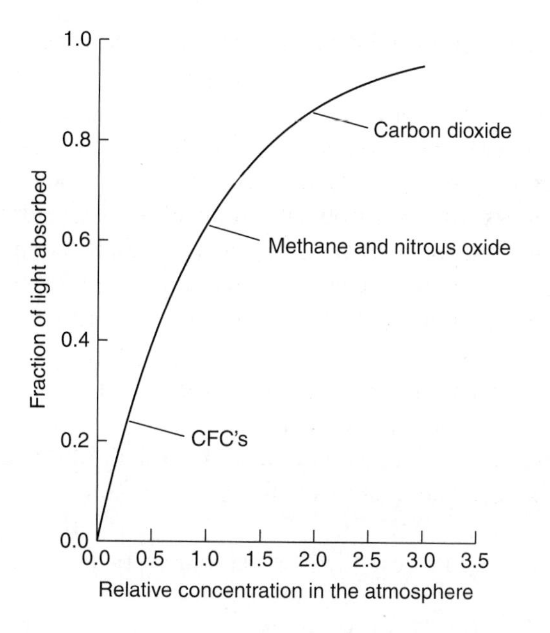

Figure 5-9 A smaller fraction of light is absorbed when the concentration of a greenhouse gas is higher. (*Based on derivation by Colin Baird in Environmental Chemistry, W.H. Freeman and Company, 1995.*)

Absorption of methane or nitrous oxide is a little more than 50 percent. Increasing their concentration has a smaller impact.

Carbon dioxide present in the atmosphere in comparatively high concentrations already absorbs close to 80 percent of the wavelengths it can absorb. Doubling the level of carbon dioxide in the atmosphere will not double its absorption but will promote a more modest increase. This is a simplified example of the types of scientific relationships among climate variables that are built into climate models. Models help to predict the future climate changes.

This is one example of a negative feedback in which nature reduces the overall impact of human-generated greenhouse gases as their concentration builds up. Climate feedbacks will be addressed in greater detail in Chapter 7.

Isotopes—Determining How Old Something Is

Researchers use isotopes in climate studies to measure the age of a sample layer in an ice core, a tree ring, or the sediment from a borehole. An application of this technique was mentioned in Chapter 2. Isotopes also can provide insight as to whether the source of carbon dioxide emissions is human or natural.

An *isotope* is a form of an element that has a different atomic structure from other isotopes of that element. What gives any element its chemical identity is the number of protons in its nucleus. For instance, carbon is carbon because—no matter what—it has six protons. Most carbon atoms also have six neutrons in the nucleus. This form of carbon, called *carbon-12* or ^{12}C, is the most common form found in nature. (So we have 6 protons plus 6 neutrons that gives a total of 12 particles in the nucleus, or carbon-12). Carbon-12 is stable. It is not radioactive, and it does not decay.

If we add 2 extra neutrons to the nucleus, there are 14 particles in the nucleus, giving us carbon-14. Carbon-14 is not stable. It decays, taking 5730 years for half of it to decay. If we have a sample of carbon-14 and half of it is gone, the sample is 5730 years old. If only one-quarter of the sample is left, we know the sample is 11,460 years old (two times the half-life of 5730 years). If only one-eighth of the sample is left, we would know that the sample is 17,190 years old (three half-lives).

Because of the two extra neutrons in the carbon-14 atoms, they are heavier. This allows them to be physically separated from other isotopes. Hydrogen can be hydrogen-1, hydrogen-2, and hydrogen-3. Oxygen also has common isotopes, namely, oxygen-16, oxygen-17, and oxygen-18 (Figure 5-10).

Isotopes are useful in determining temperature. Oxygen-18 (^{18}O), having two extra neutrons in its nucleus, is slightly heavier. During warmer conditions, more of this form of oxygen gets incorporated into ice. Figure 5-11 shows the type of relationship between the mix of isotopes and temperature.

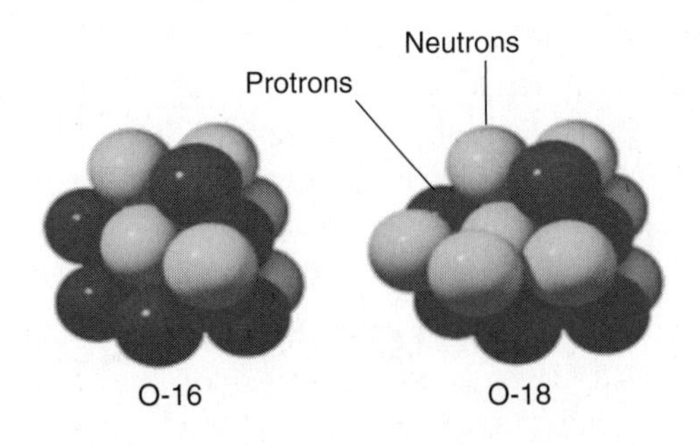

Figure 5-10 The oxygen-18 isotope has two more neutrons in its nucleus than the oxygen-16 isotope. (*Source: NOAA.*)

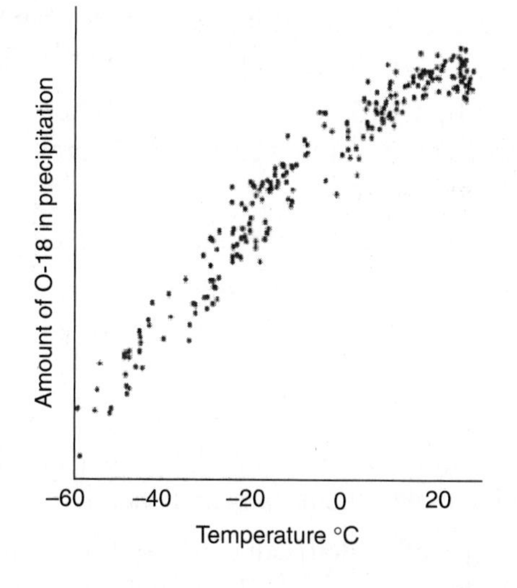

Figure 5-11 Fraction of oxygen-18 in annual precipitation versus temperature. (*Source: NOAA.*)

Key Ideas

- The atmosphere consists of the following natural gases: nitrogen (78 percent), oxygen (21 percent), and others (1 percent).
- All fuels except hydrogen have carbon and will produce carbon dioxide when burned (completely).
- Molecules of gases in the atmosphere absorb light from various parts of the electromagnetic spectrum by vibrating in ways that depends on their chemical structure.
- Nitrogen and oxygen are simple two-atom molecules. They do not absorb much of the sunlight that comes into the earth's atmosphere or the infrared light radiated from the earth's surface.
- Water (H_2O) can bend in more ways that other substances and can absorb specific wavelengths of energy from both the visible and infrared spectra.
- The greenhouse gases carbon dioxide (CO_2), methane (CH_4), and nitrous oxide (N_2O) do not absorb visible light well but are "tuned" to the infrared energy the earth radiates when heated.
- Methane is less abundant but absorbs energy 20 times more strongly than carbon dioxide.

- Ozone in the troposphere absorbs infrared energy, but because of its low abundance, it plays a relatively minor role in global warming.
- Ozone in the stratosphere absorbs ultraviolet light that has a minor cooling effect on the earth's temperature because it captures energy before it can reach the lower atmosphere.
- Halocarbons are very strongly absorbing, long-lived greenhouse gases found in the atmosphere in low concentrations.
- Industrial processes such as cement manufacturing add carbon dioxide to the atmosphere.
- Nature removes much of the new carbon dioxide that is released by human activities by absorption in the oceans. This is causing the pH of the oceans to decrease, which makes the oceans more acidic.
- Isotopes are varieties of natural elements that differ in the number of neutrons in their nuclei. Isotopes are used to determine the age of layers in ice-core samples and other fossil samples. Isotopes also can identify the origin of gases released into the atmosphere during various periods of history.
- The relative impact of greenhouse gases decreases as their concentration in the atmosphere increases.
- The integrated gasification combined cycle (IGCC) makes it easier to capture carbon from coal-fired electricity-generating plants.

Review Questions

1. What is it about methane that enables it to absorb several wavelengths of infrared light as strongly as it does?
   (a) It has strong chemical bonds.
   (b) It is a small molecule.
   (c) It can vibrate in several ways that absorb infrared wavelenghts.
   (d) It reacts with oxygen to form carbon dioxide.
2. Which of the following *most strongly* absorbs light coming directly through the atmosphere from the sun?
   (a) Nitrogen
   (b) Oxygen
   (c) Carbon dioxide
   (d) Water vapor

3. Which of the following would you expect, just looking at its chemical structure, to *most strongly* absorb light?
   (a) CO (carbon monoxide)
   (b) Cl_2 (chlorine gas)
   (c) CCl_4 (carbon tetrachloride)
   (d) O_2 (oxygen)
4. What is a by-product of making calcium oxide from calcium carbonate in the production of cement?
   (a) Nitrogen
   (b) Carbon dioxide
   (c) Water vapor
   (d) Hydrogen
5. Which of the following may store carbon dioxide in a possible carbon sequestration plan?
   (a) Methane
   (b) Calcium bicarbonate
   (c) Calcium chloride
   (d) Ozone
6. What role does ozone in the troposphere play in global warming?
   (a) Minor greenhouse gas
   (b) None
   (c) Chemically attacks carbon dioxide
   (d) Absorbs ultraviolet wavelengths
7. Which of the following produces the most carbon dioxide for a given amount of energy provided during combustion?
   (a) Coal
   (b) Oil
   (c) Natural gas
   (d) Ethanol
8. Which of the following does *not* produce carbon dioxide when it burns?
   (a) Ethanol
   (b) Biodiesel fuel

(c) Hydrogen

(d) Coal

9. If the concentration of each of the following were to double, which would cause the amount of infrared light absorbed to also double?

(a) A chlorinated, fluorinated hydrocarbon such as CF_2Cl_2

(b) Methane

(c) Nitrous oxide

(d) Carbon dioxide

10. When greenhouse gases absorb light, where does that energy go?

(a) It tightens up the chemical bonds.

(b) It is stored in the electrons of the atoms.

(c) It is trapped inside the nucleus.

(d) It causes the molecules of the gas to move.

11. Which of the following is a basic characteristic of a greenhouse gas?

(a) Absorbs visible light coming from the sun

(b) Absorbs infrared light radiated from the earth

(c) Absorbs ultraviolet light in the stratosphere

(d) Is transparent to light of all wavelengths

12. How do climatologists use isotopes?

(a) To break up ozone into elemental oxygen

(b) To store carbon dioxide

(c) To measure the concentration of carbon dioxide in an air sample

(d) To determine the age of ice-core layers

CHAPTER 6

Origin and Impact of Greenhouse Gases

The climate of the earth has gone through a recurring pattern of temperature changes which has brought the earth in and out of a series of historic ice ages. Gases in the atmosphere also have ebbed and flowed with the temperature cycles. However, the levels of greenhouse gases building up in the atmosphere today have never been this high throughout Earth's history. This chapter will explore where these greenhouse gases are coming from and how they are driving climate changes.

There are several ways we know that it is us and not nature, causing the climate to change this time. We will look into the reasons why we can say this. This chapter will examine the impact of carbon dioxide coming from burning fossil fuels, which should come as no surprise as the main culprit. We also will run down the other climate drivers, including methane, nitrous oxide, and reflectivity changes. There is a collection of unpronounceable gases that ravage the ozone layer; to add insult to injury, many of these are also greenhouse gases and will be addressed in this chapter.

We are setting out in this chapter to more precisely define the causes of climate change. This is in many ways an important prerequisite to developing a solution.

What People Add to Nature's Greenhouse

Each year human activities increase the concentrations of the greenhouse gases in the atmosphere. Figure 6-1 shows how the concentrations of these gases have increased each decade.

Carbon dioxide from the burning of fossil fuels stands out in terms of both being the largest greenhouse gas component and also increasing at the fastest rate. To enable a better comparison between the different greenhouse gases, they are listed in terms of *carbon dioxide equivalents*, which allows a comparison that includes their concentration and their impact. This will be discussed in more detail later in this chapter.

CARBON DIOXIDE

Carbon dioxide is increasing in the atmosphere. Its concentration is now over 380 parts per million (ppm) and rising rapidly. At no time in this planet's history has the carbon dioxide level ever been above 300 ppm. Today, it is at levels that are unprecedented in recent history, and it is growing at a rate that, if it continues, is likely to have a major impact on global climate.

Each year nature puts about 200 billion tons of carbon dioxide into the atmosphere, mostly from decaying plants and spewing from volcanoes. This is 30 times more than what we release into our atmosphere from our cars, power plants, and factories. However, the hundred billion or so tons of carbon dioxide that we have added to the atmosphere since 1850 are enough to initiate the climate changes that we are beginning to witness around the world. Since we add roughly 7 billion tons each year at current rates, we are continuing to tip the balance.

Nature produces and absorbs carbon dioxide, establishing what is called the *preindustrial level* of about 280 ppm in the air. This means that for every million air molecules, 280 of them are carbon dioxide. Today, the global carbon dioxide level, as a result of human activities, is nearly 100 ppm higher than the natural level. At current rates, this may reach a level of 560 ppm by the end of this century. Contributions from humans are known as *anthropogenic* greenhouse gases. Carbon dioxide is at the top of the list because of its impact on global warming. Burning fossil fuels primarily for electricity generation, transportation, industrial process heat, and heating buildings contributes carbon dioxide. Cement production stands out as an industry that makes a large contribution to global carbon dioxide levels.

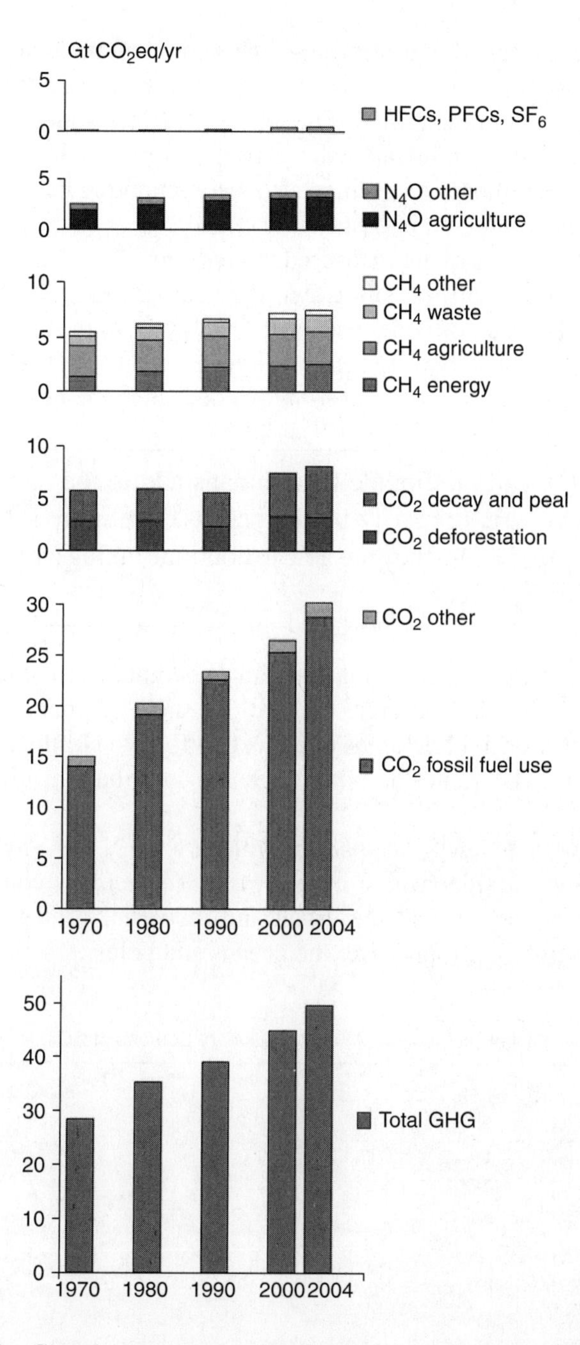

Figure 6-1 Greenhouse gas emissions broken down by type. (*Source: IPCC.*)

Forests absorb carbon dioxide through the action of photosynthesis and are considered a natural *carbon sink* because they remove carbon dioxide from the atmosphere. Beginning in the mid-1700s, settlers in North America began to clear forest land to make room for the massive westward migration that was about to take place. As the settlers removed trees, they also were removing the natural sinks, and carbon dioxide levels started climbing.

Carbon dioxide emissions are measured in *gigatons* of carbon dioxide ($GtCO_2$). The prefix *giga* means "billion." Between the years 2000 and 2005, 7.2 $GtCO_2$ (7.2 billion tons of carbon dioxide) was added each year to the atmosphere. This is an increase above the 6.4 $GtCO_2$ emissions each year from 1900 to 1999. This represents an increase of 12.5 percent from one decade to the next.

> About half the carbon dioxide that humans add to the atmosphere is removed by nature, primarily by the oceans. The more saturated the oceans become, the less they are able to continue taking carbon dioxide out of the atmosphere.

Carbon dioxide emissions come mainly from the combustion of fossil fuels. This occurs mostly in industrialized areas. Figure 6-2 show stationary sites of carbon dioxide around the world. Main areas of concentration are North America, Europe, and Southeast Asia. These sites do not include the contributions of mobile sources such as cars.

Compare this with a view from space provided by the NASA *Aqua* spacecraft in Figure 6-3. Carbon dioxide (which in this case also includes contributions from moving sources) hovers around the urban and industrial centers. Lower carbon dioxide concentrations are found over the oceans and poles.

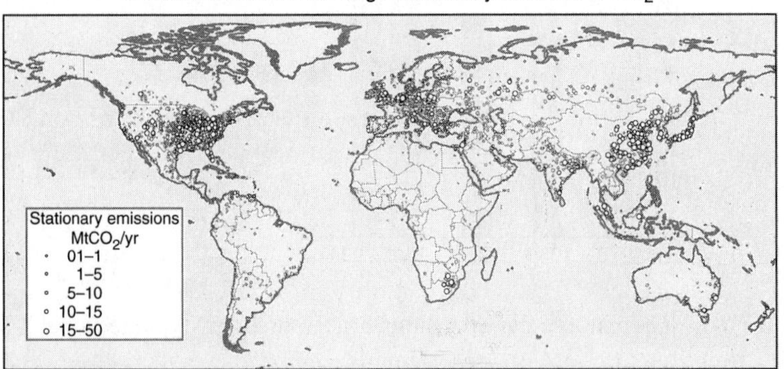

Figure 6-2 Stationary sources of carbon dioxide. (*Source: IPCC.*)

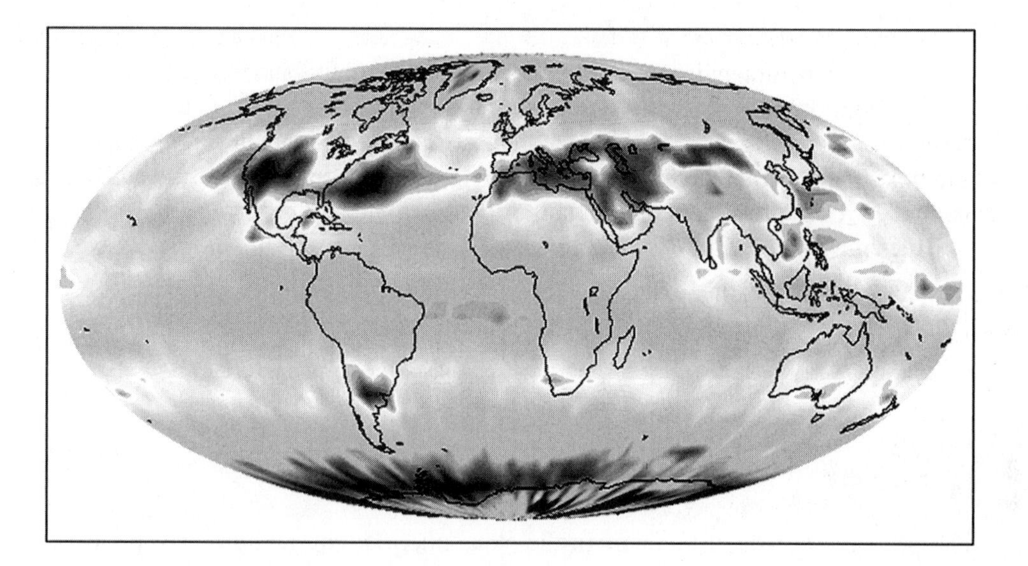

Figure 6-3 Regional carbon dioxide distribution in the mid-troposphere about 8 km (5 miles) above the surface of the earth. Areas of heaviest carbon dioxide concentration show up as darker areas. (*Source: NASA.*)

Some mixing of gases may have occurred in the atmosphere, but the highest concentrations of carbon dioxide are centered closest to their sources.

MEASURING GREENHOUSE GAS EMISSIONS

Scientists use several different ways to quantify how much of a greenhouse gas is being pumped into the atmosphere over a given time period. This is important in formulating a solution to the problem. It is also important when organizations may be assessed penalties or given credits for their greenhouse emissions. Here are three different methods for keeping track:

1. *Gigatons of carbon dioxide (GtCO_2).* This is a straightforward measure of how many (billions of) tons of carbon dioxide are emitted by a site.
2. *Gigatons of carbon (GtC).* Some people like to talk about "carbon" emissions or "carbon footprints." GtC measures the same thing as GtCO_2 but focuses only on the weight of the carbon, not the carbon dioxide. This number is 27.3 percent smaller than GtCO_2, reflecting the percentage of carbon in carbon dioxide by weight.

(*continued*)

3. *Carbon dioxide equivalents* ($GtCO_{2,eq}$). Since there are a number of greenhouse gases, to avoid dealing with each one separately, a single index is used to measure the overall effect of all the greenhouse gases having an equivalent effect as carbon dioxide. Thus, if carbon dioxide, methane, and nitrous oxide are being generated in a certain region, one number can sum up the entire impact. This is called the *carbon dioxide equivalent.*

Note: Occasionally, greenhouse emissions may be reported in kilograms or in tonnes, which is a 1000-kg unit. It is important to compare apples with apples.

METHANE

The primary sources of methane in the atmosphere are agriculture and the use of fossil fuels. The main natural sources of methane are from decomposition of organic matter in wetlands, rice paddies, and bogs. Termites are also a source of methane. Methane was once called "swamp gas" and "marsh gas" because it was commonly observed to evolve from wetland areas filled with decaying organic material such as leaves and other vegetation.

Methane is a by-product of the digestion of farm animals such as cows and pigs (in a process called *enteric fermentation*). The amount of methane depends on the diet of the animals and how the manure they produce is handled. The tendency toward larger farms favors production methods that release more methane, although the opportunity is also greater to contain it.

Methane is a main component of natural gas (the other component being ethane). Leaky natural gas pipelines inadvertently release methane, which is also released during the process of extracting natural gas or petroleum from the ground.

One molecule of methane absorbs 20 times the amount of sun energy of a molecule of carbon dioxide. However, methane is now at a level of 1774 parts per billion (ppb) in the atmosphere, which is 165 times less prevalent than carbon dioxide. (As Austin Powers might say, this is parts per *billions,* not *millions,* which in the case of concentration is much less.) For this reason, methane has less than a third of the overall impact on climate that carbon dioxide has. Studies of the relative proportion of carbon isotopes in the atmosphere indicate that the human-contributed level of methane is well above the natural level. Methane has increased from fossil fuel use, but emissions from agriculture have been stable. In contrast to carbon dioxide, the growth rate of methane emissions overall actually has declined since the 1990s. Figure 6-4 shows the increase of this greenhouse component.

The main methane sink is a photochemical process that removes methane from the atmosphere. The molecules involved are water vapor, ozone, and oxygen

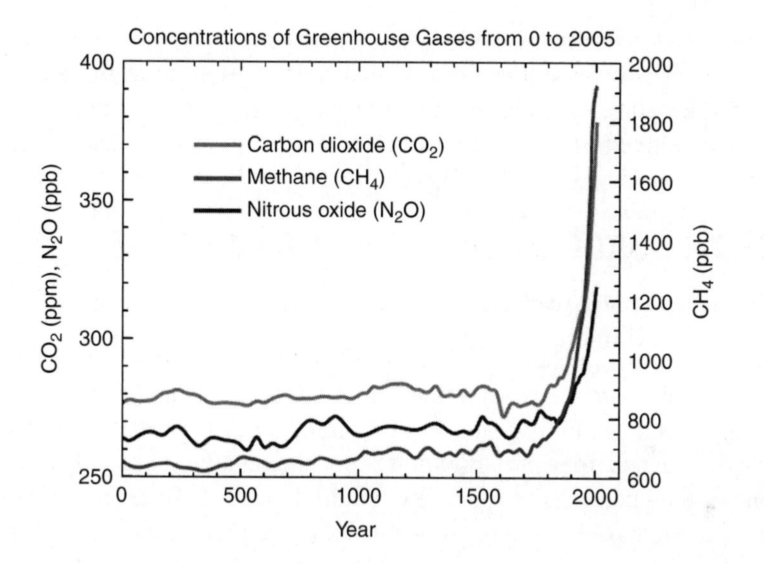

Figure 6-4 Concentration of three greenhouse gases: carbon dioxide, methane, and nitrogen dioxide. Current levels of all three are unprecedented for the past 2000 years.

interacting in the presence of light to destroy the methane molecule. Together they form a piece of a molecule called a *hydroxyl radical* (OH^-) that reacts with the methane and whose presence in the atmosphere serves as an indicator of how long methane will persist. The average lifetime for methane in the atmosphere is about 12 years, which is much less than carbon dioxide.

Methane clathrate is a slushy semifrozen mix of methane gas and ice typically found in the northern hemisphere's tundra permafrost regions and in sediment layers on the ocean floor. Some scientists have speculated that a sudden release of large amounts of methane from deposited methane clathrate called the, *clathrate gun theory,* might be a cause of abrupt past and possibly future climate changes. However, one study of ice core samples from Greenland performed by Dr. E. Brook of Oregon State University did not find a sudden increase in methane levels in past atmospheres that might have accounted for precipitous historic temperature increases.

NITROUS OXIDE

More than a third of the nitrous oxide put in the atmosphere by humans comes from the use of nitrogen-based fertilizers in agriculture. Nitrous oxide is the well-known dental anesthetic known as "laughing gas." Some nitrous oxide is formed as a result of biomass burning and certain industrial processes such as nylon production.

The main sink for nitrous oxide is a photochemical process in the stratosphere involving oxygen and resulting in a relatively long lifetime in the atmosphere of about 115–120 years. Nitrous oxide, like the other greenhouse gases, has been increasing steadily above its preindustrial values, as shown earlier in Figure 6-4.

CHLOROFLUOROCARBONS (CFCs) AND OZONE

Ultraviolet Protection

CFCs are human-made chemicals that do not exist in nature. Since they become a vapor just below room temperature and are nonflammable and nontoxic, they were used widely in refrigeration systems, insulation, and as a propellant in spray cans in the 1980s. Because they are highly stable chemically, once released into the atmosphere, they can persist for a few hundred years. Like an unwelcome house guest, they are released into the atmosphere and are pretty much there for good.

All CFCs as a group have a concentration in the atmosphere of around 1 ppb. This may seem like an insignificant presence, but it was enough to destroy ozone in the stratosphere. Stratospheric ozone protects the inhabitants of the earth by absorbing ultraviolet rays from the sun before they can enter the atmosphere. Stratospheric ozone is like nature's sunblock.

In response to concerns about ozone depletion, many governments signed the Montreal Protocol in 1989, which required the phasing out of CFCs entirely. The industrial countries were given until 1996 to do this, and the developing countries had until 2006. In terms of international cooperation, this is a success story. As a result of implementation of the Montreal Protocol, CFCs are no longer increasing in the atmosphere. Because of the very long lifetime of CFCs in the atmosphere, though, they are expected to be present near current levels for hundreds of years.

CFCs and Ozone as Greenhouse Gases

In terms of global warming, both ozone and CFCs are greenhouse gases. The presence of CFCs is partly compensated by their tendency to destroy ozone. The CFC molecules make up for their relatively small abundance by being virtual "infrared sponges," absorbing 5000–10,000 times the amount of energy from the sun as a carbon dioxide molecule. On balance, they contribute about 20 percent of the overall atmospheric heating owing to the current mix of greenhouse gases.

The CFCs were replaced by a class of chemicals called *halocarbons*. (The chemical prefix *halo* refers to the halogens, which include the elements fluorine, chlorine, bromine, and iodine.) Some of the chemical details of these gases are given in Chapter 5. These include hydrochlorofluorocarbons (HCFCs) and hydrofluorocarbons (HFCs). The HCFCs are less destructive to ozone and are preferable for that reason.

Both, however, are greenhouse gases, although they have shorter lifetimes on the order of decades rather than centuries. HCFCs are slated for being phased out by 2030.

Other compounds in this group are the perfluourocarbons (CF_4 and C_2F_6) and sulfur hexafluoride (SF_6), which are used in industrial processes. Although they are present in the atmosphere in small quantities, they have 1000-year lifetimes.

Ozone in the stratosphere absorbs ultraviolet light, preventing it from being absorbed closer to the earth's surface. Some of this can be readily radiated back into space. Thus ozone in the stratosphere actually has a slight cooling effect on the earth's overall energy balance.

AEROSOLS—AN UMBRELLA ABOVE THE GREENHOUSE

Human activities also add solid suspended material—called *aerosols*—to the atmosphere. These include suspended carbon (soot and ash), carbon in compounds, dust, and oxides of sulfur and nitrogen. Together, the aerosols produce a cooling effect by reflecting incoming sunlight back into space before it has a chance to enter the atmosphere. Overall, the aerosols act to restrict some of the sunlight from heating the earth. In some cases they absorb some of the incoming light. Aerosols also can influence cloud formation and precipitation.

How Do We Know Where the Greenhouse Gases Come From? Human Fingerprints versus Nature's Pawprints

There is no question that carbon dioxide is a natural part of the atmosphere. In the past, nature has pumped a lot of carbon dioxide into the atmosphere during the warm interglacial periods between the frozen depths of the ice ages. Volcanoes may have released this carbon dioxide, or as some scientists have suggested, it may have been released from the oceans as they warmed up from other causes (such as changes in the earth's orbit). How do we know where the carbon dioxide is coming from?

One tool that climatologists use is *isotopes* (which were introduced in Chapter 5). Fossil fuels are literally fossil remnants of organic material left in the ground during the *Carboniferous Period* between 360 and 286 million years ago. Fossil fuels, having been around for so long, are depleted of any traces of carbon-14, which has a half-life of 5730 years. After 10 or more half-lives, or 57,300 years, the small amount of naturally occurring carbon-14 would have decayed to an insignificant amount. By comparing the amount of carbon-14 in carbon dioxide in the air, scientists are able to recognize the telltale signature of fossil fuels as the source of these emissions.

This result is confirmed by carefully keeping track of how much fossil fuel has been combusted and how much land has been cleared and relating this to the amount of carbon dioxide that is stored in the various sinks, such as the oceans, forests, and soils. The 500 billion tons of carbon dioxide that were released into the atmosphere during the industrial period would have been enough to raise the level in the atmosphere to 500 ppm. Currently, the carbon dioxide level is roughly 380 ppm because some of the excess carbon dioxide has been absorbed. This carbon dioxide level is consistent with the amount known to have been added.

Combustion of carbon-containing fuels consumes oxygen. Another method used to identify the source of the carbon dioxide level is to measure the oxygen level in the atmosphere. Scientists are able to detect a decrease in the percentage of oxygen in the atmosphere consistent with the amount needed to burn the fossil fuels. If ocean warming were responsible for the increase in atmospheric carbon dioxide, there would have been an increase rather than a decrease in the oxygen level in the atmosphere.

More carbon dioxide is released into the atmosphere in the northern hemisphere than in the southern hemisphere. We saw this earlier in Figures 6-2 and 6-3. Data comparing the difference between readings in the northern and southern hemispheres are shown in Figure 6-5 and serve as further proof that the increased carbon dioxide levels in the air are the result of people.

The more carbon dioxide is pumped into the atmosphere, the bigger is the difference. How is this possible? The reason is that more carbon dioxide is coming from the northern hemisphere because that is where the earth is most industrialized and that is where most of the fossil fuels are being burned.

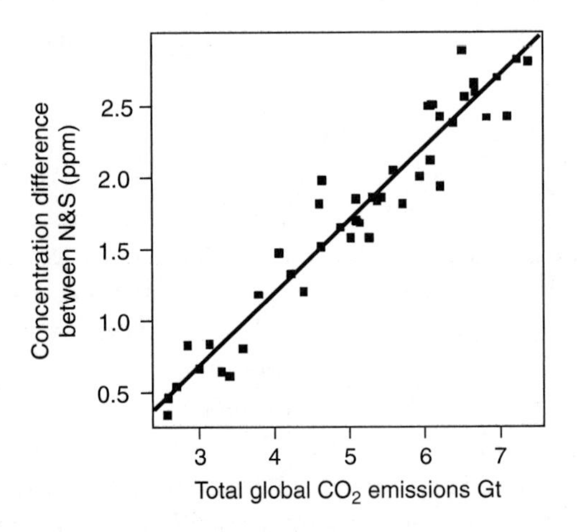

Figure 6-5 Difference in carbon dioxide levels between the northern and southern hemispheres as overall concentration increases. The greater amount of carbon dioxide in the northern hemisphere is the result of industrialization. (*Source: IPCC AR4 Working Group 1.*)

How the Greenhouse Gases Affect the Atmosphere

ABILITY TO ABSORB SUNLIGHT

We have seen that there is a natural greenhouse effect that keeps the earth's temperature in balance and without which the earth would be a lifeless ball of ice. As we saw in Chapter 3, without the natural greenhouse effect, the earth would be at a far less habitable temperature of around –19°C (–2°F). Some molecules absorb energy more readily than others. Carbon dioxide is not the strongest absorber, but because it is present in such large quantities, it dominates the rest. Methane and nitrous oxide absorb energy more effectively, whereas the fluorine-bearing compounds are "energy sponges." Fortunately, for now, they are present in the atmosphere at only trace levels, but their potential to contribute to global warming is enormous (Table 6-1).

RESIDENCE TIMES—HOW LONG THE GREENHOUSE GASES LAST

The long-term impact of greenhouse gases depends on, among other things, how long they persist in the atmosphere. Methane is attacked by other chemicals in the air and is removed fairly quickly. The CFCs are very stable and are likely to be in the atmosphere for the long haul. Unlike other greenhouse gases, carbon dioxide does not break down to any great extent in the atmosphere. Rather, it is removed

Table 6-1 Ability of Various Greenhouse Gases to Absorb Energy Compared with Carbon Dioxide

Gas	Ability to Absorb Infrared Energy Compared with Carbon Dioxide
Carbon dioxide (CO_2)	1
Methane (CH_4)	26×
Nitrous oxide (N_2O)	216×
CCl_2F (a chlorofluorocarbon, CFC)	22,860×
CH_3CF_3 (a hydrofluorocarbon, HFC)	9,290×
Sulfur hexafluoride (SF_6)	37,140×

Source: IPCC AR4 Working Group 1 Technical Summary, 2007.

Table 6-2 Lifetime of Greenhouse Gases in the Atmosphere

Gas	Lifetime in the Atmosphere
Carbon dioxide (CO_2)	100 years
Methane (CH_4)	12 years
Nitrous oxide (N_2O)	115 years
Chlorinated fluorinated hydrocarbons (CFCs)	45–1700 years

from the atmosphere by the natural carbon cycle. A roughly 100-year time span is assigned to carbon dioxide so that it can be compared with other greenhouse gases (Table 6-2).

GLOBAL WARMING POTENTIAL

The *global warming potential* (GWP) of each gas gives an indication of how effective a given amount of that gas is in contributing to global warming. This combines the ability of the gas to absorb infrared radiation with how long it survives in the atmosphere. A gas that absorbs very strongly but for a short time may have less of an impact than a gas that absorbs weakly but has a longer lifetime. All global warming potentials are compared with carbon dioxide, which is defined to have a GWP of 1 throughout its residence time in the atmosphere. GWPs typical of the various groups of gases in the atmosphere are listed in Table 6-3. As a greenhouse gas is removed from the atmosphere, its impact decreases. For this

Table 6-3 Global Warming Potential for a Sample of Greenhouse Gases

Gas	Lifetime (years)	GWP for Given Time Horizon		
		20 Years	100 Years	500 Years
Carbon dioxide (CO_2)	100	1	1	1
Methane (CH_4)	12	72	25	7.6
Nitrous oxide (N_2O)	114	289	298	153
CCl_2F (a chlorofluorocarbon, CFC)	13.8	3300	1,300	400
CH_3CF_3 (a hydrofluorocarbon, HFC)	52	5,890	4,470	1,590
Sulfur hexafluoride (SF_6)	3,200	16,300	22,800	32,600

Source: IPCC AR4 Working Group 1 Technical Summary, 2007.

reason, the GWP depends on whether we are talking about it impact after 20, 100, or 500 years.

The greenhouse gas components methane and nitrous oxide have higher GWPs than carbon dioxide. However, because they are present in lower concentrations than carbon dioxide in the atmosphere, they will have a smaller overall contribution to global warming. The last three listings in the table are fluorine-containing compounds. Although currently present in the atmosphere at very low concentrations, they can be very dangerous because of their extremely long lifetimes and ability to contribute to global warming for many years.

> Greenhouse gases containing chlorine and fluorine are present at low levels in the atmosphere and are currently playing a relatively minor role in effecting global temperatures. However, like an unwanted house guest, they tend to stay around for a long time and absorb heat as if they were energy sponges. If allowed to build up in the atmosphere, these gases could become a far more serious threat.

OVERALL IMPACT—RADIATIVE FORCING

Scientists have developed a way to compare the various factors affecting the earth's temperature based on the idea of *radiative forcing*. We know that the sun delivers solar power at a rate of 1370 W for every square meter of surface. The total impact of human activities adds 1.6 W/m^2 to this value. A positive forcing tends to make the earth warmer. A negative forcing offsets the positive factors and contributes a cooling effect.

According to the Intergovernmental Panel on Climate Change (IPCC), the largest single contributor is carbon dioxide added to the atmosphere by human activities such as driving cars or using coal to produce electricity. Carbon dioxide represents half the impact of all the greenhouse gases. The other greenhouse gases (i.e., methane, nitrous oxide, and the various chlorinated halocarbons), which are present in much lower concentrations but which absorb infrared light very strongly, contribute about an additional third. Ozone plays a mixed role. In the troposphere, ozone causes warming. High in the stratosphere, however, ozone absorbs incoming light before it has a chance to warm the atmosphere.

Global dimming is alive and well. Particles in the air and clouds enhanced by pollution reflect sunlight before it can enter the atmosphere. As a result of particles in the air that reflect sunlight back into space, the earth is a little cooler than it otherwise would have been.

Overall, the combined effects of forcing influences result in a net increase in atmospheric temperature.

RADIATIVE FORCING—LIKE A DIMMER SWITCH FOR INCOMING SOLAR ENERGY

The term *radiative forcing* describes itself very well. *Radiative* refers to its effect on the overall energy being received from solar radiation. *Forcing* implies how a particular influence affects climate. A good way to think of radiative forcing is that each forcing effectively either turns the sun up a little brighter or (in the case of negative effects) turns it down to make it dimmer. The greenhouse gas components methane and nitrous oxide have higher global warming potential than carbon dioxide. However, because they are present in lower concentrations than carbon dioxide in the atmosphere, they have a lower radiative forcing.

FACTORS THAT HEAT THE EARTH

The following chart shows radiative forcings identified in the IPCC report, *Climate Change 2007.* Both positive forcings, which increase the temperature of the earth, and negative forcings, which decrease the temperature of the earth, are shown. Overall, these combine to produce a net heating effect of +1.6 W/m^2. All forcings shown in the chart are human, with the exception of the forcing resulting from variations in solar intensity.

Radiative Forcings	
Carbon dioxide	+1.66 W/m^2
Methane (including stratospheric water)	+0.55 W/m^2
Ozone (in the troposphere)	+0.35 W/m^2
Halocarbons	+0.34 W/m^2
Nitrous oxide	+0.16 W/m^2
Contrails	+0.01 W/m^2
Natural Factors Contributing to Heating	
Increase in solar intensity	+0.12 W/m^2
Human Factors that Cool or Offset the Heating Effects	
Aerosols (direct and increased clouds)	–1.2 W/m^2
Stratospheric ozone	–0.05 W/m^2
Surface reflectance change	–0.1 W/m^2
Overall net forcing from all sources	+1.6 W/m^2

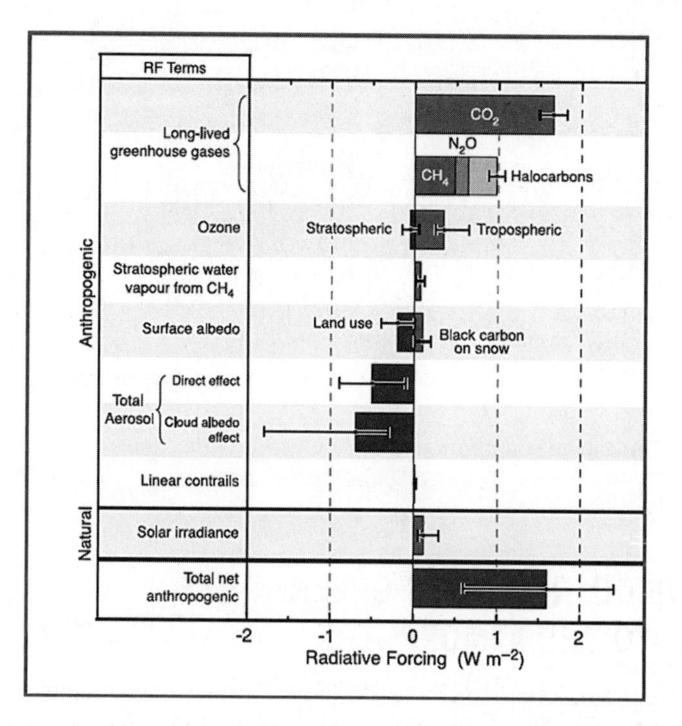

Figure 6-6 Radiative forcings. (*Source: IPCC.*)

The radiative forcings from all sources are shown in Figure 6-6. Both positive and negative effects are shown, with a net overall heating effect.

Where the Human-Added Greenhouse Gases Come From

Topping the list is carbon dioxide from fossil fuel use, comprising 57 percent of all greenhouse gases. This percentage becomes much higher as countries become industrialized. Just over 19 percent of the greenhouse gases is carbon dioxide contributed by decay, following forest clearing, forest and brush fires, and other activities related to land use. All told, carbon dioxide accounts for nearly 80 percent of the greenhouse gases added to the earth's atmosphere. Methane, nitrous oxide, and halocarbons make up the balance. This is summarized in Figure 6-7.

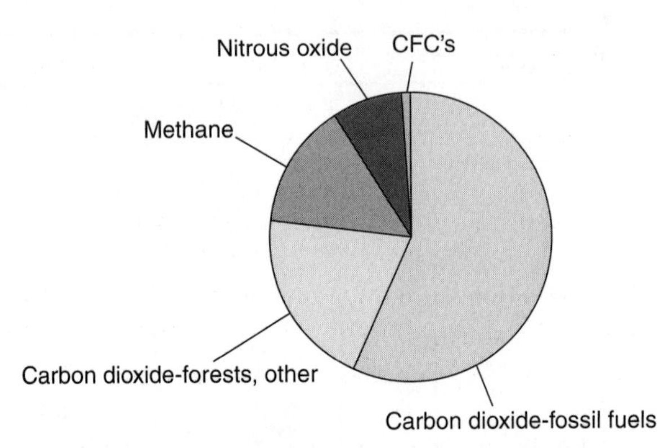

Figure 6-7 Percentage contributions from the various greenhouse gas emissions in 2004. (*Source: IPCC.*)

WHAT WE DO THAT PUTS GREENHOUSE GASES INTO THE ATMOSPHERE

Providing energy to the world produces the largest amount of greenhouse gas emissions. This consists of nearly 26 percent of the total, as shown in Figure 6-8. Running industry around the world is next, adding 19.4 percent. Other human activities, such as forestry (17.4 percent), agriculture (13.5 percent), transportation (13.1 percent), heating or cooling buildings (7.9 percent), and landfills (2.8 percent), also contribute their share. The mix for the more industrialized countries will have greater proportions from the energy and transportation areas than the world as a whole.

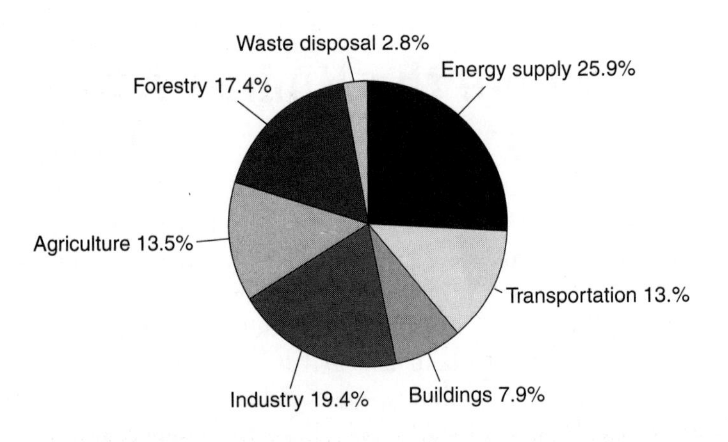

Figure 6-8 Greenhouse gas emissions by sector.

> The two largest contributors of greenhouse gases are burning *coal* to produce electricity and burning *petroleum* products to move vehicles.

DEVELOPED VERSUS EMERGING ECONOMIES

The United States produces a larger amount of greenhouse gases than any other country. With around 5 percent of the world's population, the United States contributes roughly 25 percent of the world's greenhouse gases, making the United States one of the highest per-capita contributors to global warming. This also means that the rest of the world contributes 75 percent of the greenhouse gas emissions.

China is the fastest growing contributor to global warming and is on pace to surpass the United States during this decade. As a rapidly developing nation, China currently has a far lower per-capita greenhouse gas contribution. The European Union contributes nearly 14 percent of the world's greenhouse gases.

The mature economies have developed transportation and electricity-generation systems. The emerging economies are rapidly assembling the infrastructure to provide an improved standard of living for their citizens. This means electricity and cars—and the associated global warming they bring. There are many more people in countries with emerging economies. If they use the same greenhouse emission–producing forms of energy as the present industrialized countries have, the pace of global warming will accelerate. Some forecasters expect that by 2015, the total contributions from the emerging economies will overtake those of the more developed countries. Current trends are shown in Figure 6-9.

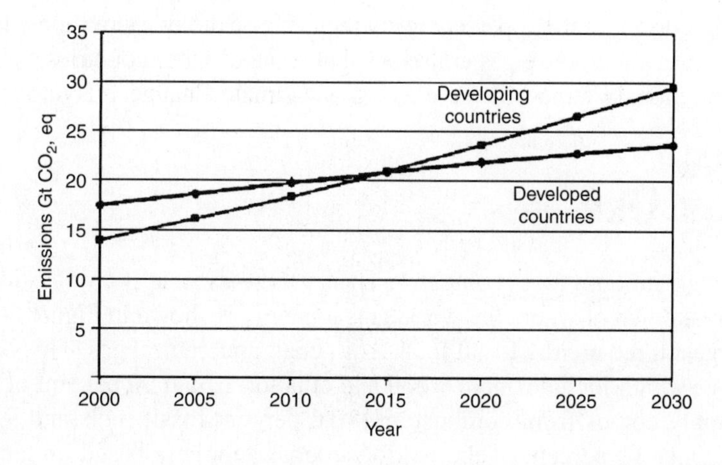

Figure 6-9 Comparison of greenhouse gas emissions by emerging versus developed countries. (*Source: EPA.*)

Table 6-4 Carbon Dioxide Contributions of Various Counties Overall and for the Year 2005

Country	Total CO_2 Emissions (%)	2005 CO_2 Emissions (%)
United States	27.8	20.5
China	7.8	18.0
Russia	7.5	5.4
Germany	6.7	2.7
United Kingdom	6.1	2.0
Japan	3.9	4.4
India	2.4	4.9
Rest of Europe	18.3	12.4
Rest of the world	15.5	21.7
Ships/airplanes	4	5

Source: Dangerous human-made interference with climate: a GISS modelE study., J Hansen and 46 co-authors (2007) Atmos. Chem. Phys., 7, 2287-2312, PDF available at http://www.columbia.edu/~jeh1/canweavert.pdf]

It is clear from looking at Table 6-4 that no single country is responsible for the world's greenhouse gases and no country alone can correct the problem independently. The United States has contributed the most to the overall buildup of carbon dioxide in the atmosphere to date. It is contributing a smaller percentage in 2005, not because it has reduced it carbon dioxide emissions but rather because the rest of the world is catching up. China is becoming a close second and is closing rapidly. Some countries are showing a decrease in the percentage of their carbon dioxide emissions. This shows progress toward greenhouse gas emissions that some of these countries committed to at The International Framework Convention on Climate Change in Kyoto in 1997.

World Energy Use

Most of the greenhouse gases come from energy use—especially combustion of fossil fuels. A breakdown of where the world gets it energy is shown in Figure 6-10.

The largest three are the fossil fuels—oil, coal, and natural gas. This is followed by biomass, which includes both wood and ethanol. About 90 percent of the world energy supply comes from combustion (80.6 percent fossil fuels and 9.4 percent biomass). Only 10 percent of the world's energy supply is based on technologies that do not produce greenhouse gases. These include nuclear energy, hydroelectric power, and minor contributions from emerging renewables (Figure 6-11).

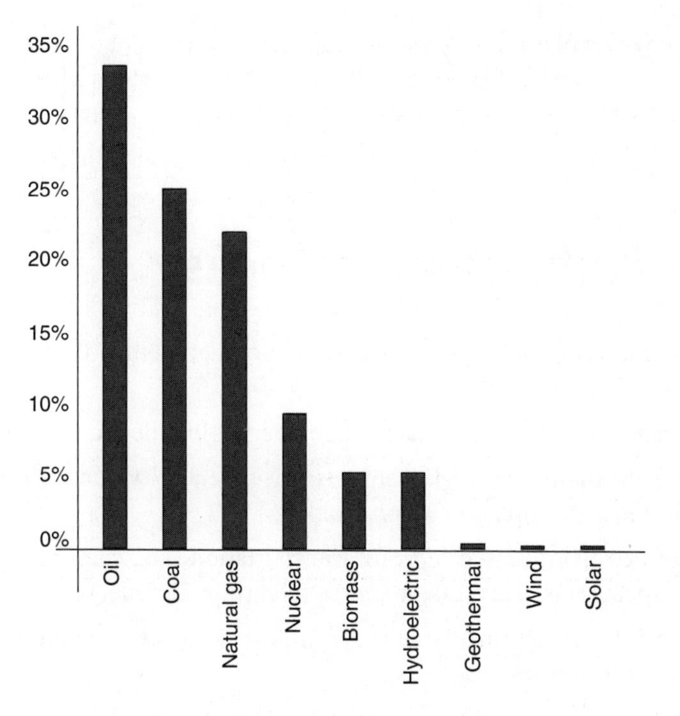

Figure 6-10 Fuels used to produce the world's energy.

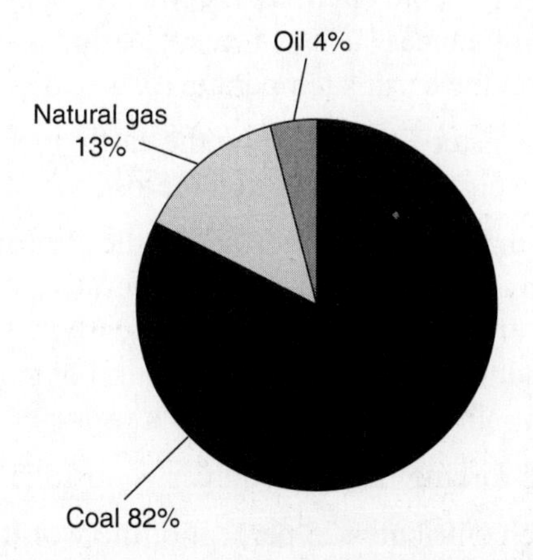

Figure 6-11 Carbon dioxide emissions from U.S. electricity generation. (*Source: Energy Information Administration.*)

Coal-fired power plants are dominant, providing 82 percent of the carbon dioxide generated in the process of producing the United States' electricity. Natural gas is the only other significant contributor to electricity generation. This is not surprising given the abundance of coal in the United States.

Defining the Problem—Key Ideas

- The main greenhouse gases are carbon dioxide, methane, nitrous oxide, and CFCs.
- The major contributors of carbon dioxide worldwide are fossil fuels.
- Coal is the main source of energy used for electricity generation worldwide and is a major source of carbon dioxide.
- Use of petroleum-based fuels for transportation (i.e., cars, buses, trucks, and airplanes) is also a major source of carbon dioxide.
- The major source of methane is agricultural practices and production of fossil fuel products.
- Nitrous oxide comes primarily from agricultural sources.
- CFCs used throughout industry are minor greenhouse gases at present but have an enormously long lifetime in the atmosphere.
- Radiative forcing is a concept used to compare the impact that greenhouse gases and other influences such as forest clearing and solar intensity variations have on the earth's temperature.
- Several items have a cooling affect on the earth, including reflectance of incoming solar energy by air pollution (aerosols).
- Scientists know that carbon dioxide added the atmosphere comes from human sources by measuring the mix of naturally occurring elements that give a signature of its origin. These include carbon-14, carbon-13, and carbon-12. In addition, the ratio of distribution of carbon dioxide between the northern and southern hemispheres is a factor.
- Overall radiative forcing from all sources is positive.
- The United State contributes 25 percent of the world's greenhouse gases.
- The rest of the world generates 75 percent of the greenhouse gases
- Rapidly developing countries are increasing their greenhouse gas releases. By 2015, developing countries are expected to produce more greenhouse gas emissions than industrialized countries. China is very close to overtaking the United States as the largest contributor of greenhouse gases.

Review Questions

1. Which fossil fuel currently accounts for the largest share of the world's energy production?
   (a) Coal
   (b) Oil (petroleum)
   (c) Natural gas
   (d) Methane
2. Which of the following *best describes* the concentration of methane in the atmosphere?
   (a) Increasing slowly
   (b) Increasing rapidly
   (c) Stabilizing and starting to decline
   (d) Completely eliminated from the atmosphere
3. Most electricity in the world is produced by which source of energy?
   (a) Coal
   (b) Oil
   (c) Natural gas
   (d) Nuclear
4. Which of these renewable sources of energy contributes the largest share of the world's electrical energy?
   (a) Wind
   (b) Solar
   (c) Geothermal
   (d) hydroelectric
5. What kind of radiative forcing do aerosols have?
   (a) Contributes to cooling
   (b) Contributes to warming
   (c) No effect
   (d) Depends on their location
6. What is the effect on climate of absorption of sunlight by stratospheric ozone?
   (a) No effect
   (b) Contributes to warming of the troposphere

(c) Contributes to cooling of the troposphere

(d) Contributes to cooling of the stratosphere

7. What percent of the world's energy comes from a source *other than* the combustion of fossil fuels?

   (a) 1 percent

   (b) 10 percent

   (c) 40 percent

   (d) 80 percent

8. Most of the electricity produced in the United States comes from

   (a) nuclear power plants.

   (b) biomass-burning plants.

   (c) hydroelectric plants.

   (d) coal-burning plants.

9. By what year are the emerging economies of the world expected to produce more greenhouse gas emissions than the present group of industrialized nations?

   (a) 2009

   (b) 2015

   (c) 2040

   (d) 2080

10. Which of the following is the *least significant* positive contributor to radiative forcing?

    (a) Contrails

    (b) Increase in solar intensity

    (c) Absorption by nitrous oxide

    (d) Absorption by ozone in the troposphere

PART THREE

What We Can Expect and What We Can Do

CHAPTER 7

Consequences of Global Warming

We are now ready to link cause and effect. This chapter investigates how much a particular change in a greenhouse gas concentration or surface reflectivity affects climate. We will explore climate models, which are computer programs that give climate scientists a way to predict what may happen next. The wild card, of course, is guessing what people, companies, and governments around the world may choose to do. Various possibilities are explored to better understand the impact of different efforts. This chapter will present forecasts based on different scenarios. In Chapter 8 we will see what solutions are available.

Climate Models

WHAT CLIMATE MODELS DO

Climatologists try to predict what a particular set of changes will do to the earth's climate. How hot will the air get? How much ice will melt next year in Greenland? How much will sea level rise off the coast of Bangladesh in 200 years?

Unlike most other scientists, climatologists cannot just go out and do an experiment as a chemist, physicist, or biologist might do. The use of models provides climatologists with a method to test their theories that is common to other sciences as well. Climatologists then can begin to call their shots by predicting the outcome of a set of climate conditions. They then compare their predictions with actual results to validate their method and to fine-tune their models. Where models overlap, climate data derived from historical records strengthens the correlation between efforts to understand the future and investigations focused on the past.

The Intergovernmental Panel on Climate Change (IPCC) expects that climate models will be able to reliably predict changes in critical climate variables. The most repeatable results will be on large (continental) scales. Differences between models depend mostly on different estimates of the effect of feedbacks, as well as on how periodic climate patterns such as the El Niño southern oscillation (ENSO) are handled in different ways in different models. As models are tested, some of these differences are expected to be reconciled.

WHAT GOES INTO A CLIMATE MODEL?

The basic tool in the climatologist's toolkit for making predictions is the *climate model*. A *model* is a mathematical description that relates the physical, chemical, and biologic properties of a system. Models address cause-and-effect relationships and include the impact of feedback.

Models have varying degrees of complexity and can include the following elements:

1. Initial physical conditions are established, such as solar intensity, starting temperatures of air and water, salinity, greenhouse gas concentrations, absorption properties for those gases, and albedos of all exposed surfaces.
2. Cause-and-effect relationships between the variables, including direct forcing and feedback, are defined. This includes equations that define the energy balance, climate sensitivity assumptions (temperature changes for given greenhouse gas concentration increases), and ocean heat uptake.
3. Heat transfer between vertical layers as a result of either radiation or convection is included. This requires definition of boundaries between the layers, gradients within those layers, and mixing that occurs.
4. Horizontal interactions are added. Simpler models deal with continental-scale building blocks, whereas more sophisticated models include greater resolution both horizontally and vertically. Greater computing power usually is needed as more granularity is introduced into the models.

5. Dynamics of atmospheric and ocean circulations are added. Biochemical cycles (such as enhanced plant or algae growth) are represented.
6. Modeling aerosols and clouds and how they interact and initiate precipitation remains an area that is currently being refined.
7. Models then investigate various scenarios. Scenarios represent the various results that could be seen depending on what changes to greenhouse gas emissions are put in place around the world.

The various scenarios used are listed in Table 7-1. As expected, the scenarios that emphasize a transition to nonfossil fuels result in temperature spreads centered on a lower mean. The scenarios that are more "business as usual" are the ones showing the highest projected temperature ranges. Table 7-2 (later in this chapter) shows how much of a difference a few degrees can make.

FEEDBACK

Climate feedback occurs when an initial change triggers a second change that, in turn, influences the first process. The impact of a change such as an increase in a greenhouse gas level can be greater or smaller depending on other factors that may or may not be brought into play. Feedback can be either positive or negative.

Positive feedback intensifies the overall impact of the original change. An example is a public address (PA) system where some of the amplified sound gets picked up by the microphone and then get amplified again until you get the squealing sound known as *feedback*. Positive feedback tends to be destabilizing in the sense that it leads to more of a runaway situation than equilibrium.

Negative feedback occurs when other factors reduce the impact of a change. Negative feedback tends to create stability. An example of negative feedback is the thermostat in your house. As the temperature drops, the thermostat responds by

Table 7-1 Description of Various Climate Scenarios

Scenario	Characteristic
A1T	Rapid growth— reliance on nonfossil fuels
A1B	Rapid growth—balance between fossil fuels and alternative sources
A1F1	Rapid growth—continued reliance on fossil fuels
B1	Transition to service/information economy with clean technologies
B2	Environment protection mostly at local levels
A2	Rapid growth—less coordination

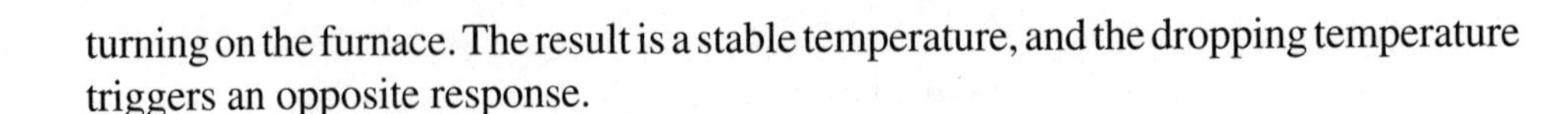

turning on the furnace. The result is a stable temperature, and the dropping temperature triggers an opposite response.

Examples of Positive Climate Feedback

Melting. Greenhouse gases cause an increase in atmospheric temperature. The increased temperature causes melting of snow on glaciers and ice caps. As the snow and ice melt, they become less reflective and retain more heat because the darker surface reflects less sunlight. As a result of this feedback effect, the atmosphere gets even warmer.

Water vapor. Warmer temperatures increase evaporation. Water vapor in the air is a greenhouse gas and will absorb greater amounts of sunlight. Because of this absorption of sunlight, the atmosphere gets even warmer.

Decay of biomass. Increased temperatures result in drought. Plants die and decay. The decaying plant material releases even more carbon dioxide to the atmosphere. With more carbon dioxide in the atmosphere to absorb more infrared radiation, the atmosphere becomes even warmer.

Forests replacing tundra. Because of global warming, tundra is melting, and forests are able to grow at higher latitudes. Tundra has a much higher surface reflectivity than forests. Thus the poleward movement of forests—caused by higher temperatures—contributes to increasingly higher temperatures.

Release of methane from permafrost. Warmer climates melt permafrost in North America, Europe, and Asia. Methane has been trapped in the tundra in the form of methane clathrates for many centuries. As methane is released into the atmosphere, it is able to function as a greenhouse gas and perpetuate the cycle of warming.

Carbon dioxide dissolved in the oceans. Higher ocean temperatures reduce the solubility of carbon dioxide in the oceans. Since less carbon dioxide is removed to the oceans, more is available to function as a greenhouse gas. With more carbon dioxide in the atmosphere, the atmosphere gets warmer than it otherwise would have got.

Glacier descent. As global warming occurs, glaciers melt and descend their mountain slopes. As the glacial mass moves to lower altitudes, the atmospheric temperature increases. The further down the slope the glaciers go, the warmer they get. The warmer temperatures toward the bottom of the slope accelerate the melting process.

Destabilization of glaciers. The ends of glaciers (tongues) reach a point where they separate from the glacier and break off into the sea. This has occurred throughout Greenland and has been observed in the Larsen ice shelves in Antarctica. Scientists have noticed that without the end pieces that hold back the river of slush coming after a glacier, the rest of the glacier moves more quickly. This is a positive feedback whose impact appears to be greater than initial expectations.

Examples of Negative Climate Feedback

Clouds. Warmer temperatures promote increased evaporation. Evaporation of water leads to the formation of clouds. Clouds typically reduce the amount of sunlight that can reach the surface of the earth. With less sunlight striking the earth's surface, the earth tends to get cooler.

Photosynthesis. As the atmospheric temperature goes up, the amount of carbon dioxide in the atmosphere also goes up. Increasing carbon dioxide in the air promotes plant growth. Plants help to remove additional carbon dioxide from the air, which counters the initial warming effect.

Infrared radiation. The warmer the atmosphere, the more effectively it radiates infrared radiation to space. As the atmosphere emits infrared radiation better, the earth does not get as hot as it might have. The process of atmospheric radiation reduces the original impact of the sun's energy through a negative feedback.

Glaciers melting. As global warming progresses, glaciers release larger amounts of freshwater into the North Atlantic. The decreased salinity slows the thermohaline circulation (Gulf stream). Less heat is moved to the North Atlantic, leading to an overall (local) cooling effect.

Snow buildup on glaciers. Warmer temperatures cause increased moisture content in the air. The elevated moisture levels will result in greater snowfall and a buildup of the ice pack in the interior regions of Greenland and Antarctica. This is a negative-feedback effect because the higher temperatures lead to a buildup of the massive ice sheets (in addition to melting).

STABILIZATION AND COMMITMENT

Stabilization

Let's say that some changes are made around the world and emissions are held steady at some new level. Climate models evaluate what will happen at that new stabilization level.

- The model can evaluate an *equilibrium* level that indicates what will happen to the climate after an extended period of time.
- The model also can evaluate a *transient* level that determines what will happen after a specific time period, typically 20 years.

Figure 7-1 show the results of climate models that predict the equilibrium temperature increase above preindustrial levels for various greenhouse gas levels. Just so that we do not have to worry too much about which greenhouse gases are causing the temperature increase, the concentrations are given as carbon dioxide equivalents (CO_2, eq). The center line gives the best estimate. The top line gives the highest range, and the bottom line gives the lower range. For instance, if the greenhouse gas level rises to 550 parts per million (ppm) concentration, the best estimate for the average global temperature would be just under 3°C (5.4°F). Notice that this graph is not linear. As the carbon dioxide concentration builds up, there is less infrared radiation left for the added carbon dioxide to absorb. (This relationship is described in Figure 5-9.)

Results like these from climate models are used to predict the consequences of greenhouse gases added to the atmosphere. Scientists modeling climate have set up several scenarios that enable them to study “what ifs.” These can include different conditions relating to population growth, economic development, and programs implemented in various places to rein in greenhouse gas emissions. The various results are valid only to the extent that the assumptions on which they are based are realistic, and there is not necessarily a single correct answer. By looking at a wide range of possibilities, it is possible to establish a more comprehensive overall picture.

Commitment

The earth’s climate cannot stop on a dime. If all greenhouse gas emissions were suddenly and magically frozen and held at today’s levels without increasing, the

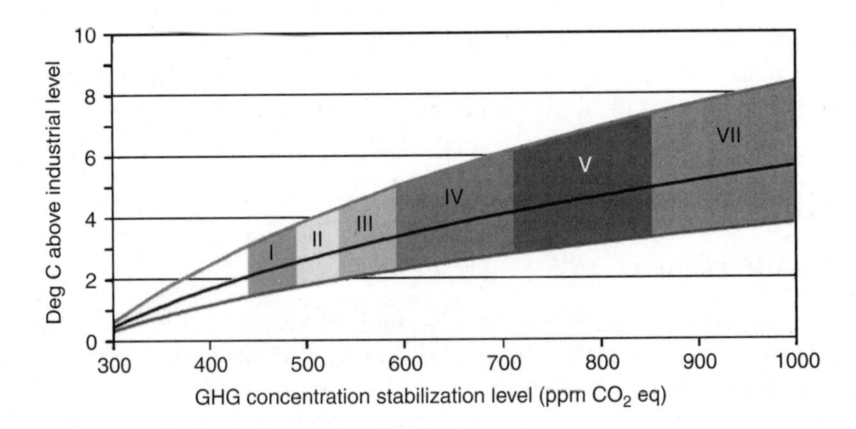

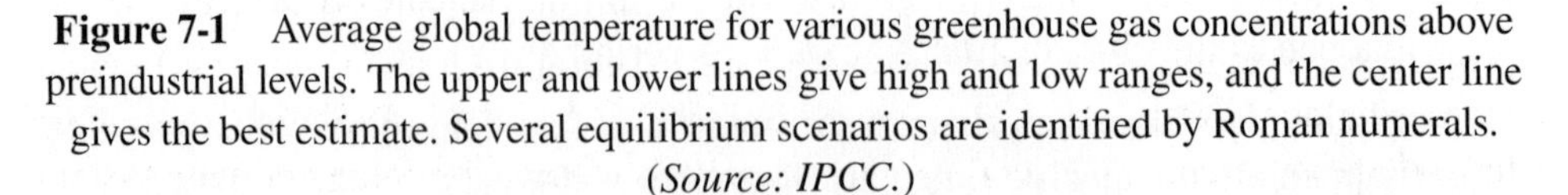
Figure 7-1 Average global temperature for various greenhouse gas concentrations above preindustrial levels. The upper and lower lines give high and low ranges, and the center line gives the best estimate. Several equilibrium scenarios are identified by Roman numerals. (*Source: IPCC.*)

earth would continue to warm up because it still would be absorbing more heat than it did in preindustrial times. Even if all greenhouse gas emissions were magically rolled back to preindustrial levels today, the earth still would warm up because of the increased concentration of greenhouse gases that have accumulated in the atmosphere for the past several decades.

For example, if greenhouse gas stabilization occurred in 2100 (for two reasonably likely IPCC scenarios), an additional increase of 0.5°C (0.9°F) still would be expected by 2200. For this case, thermal expansion alone (not even including the effects of snow and ice melting) would lead to an average global sea level increase of 0.3–0.8 m (1–2.6 ft). We have committed ourselves to a certain unavoidable level of global warming, the severity of which will depend on the concentration of greenhouse gases at which the earth stabilizes.

More than half the carbon dioxide added to the atmosphere is removed within about a hundred years. Some smaller fraction, about 20 percent, will take thousands of years before it is removed naturally. For this reason, carbon dioxide levels in the atmosphere will continue to build up even if emission levels are reduced. Small reductions of carbon dioxide emissions (on the order of 10–30 percent) could be expected to result in a proportional decrease in the growth rate. However, even if all carbon dioxide sources were cut off immediately, the carbon dioxide concentration in the atmosphere would be expected to decline only 40 ppm throughout the twenty-first century. Methane and nitrous oxide have shorter residence times, whereas chlorofluorocarbons (CFCs) have much longer residence times.

How Much and How Soon Will Climate Change?

SENSITIVITY—WHAT HAPPENS IF CARBON DIOXIDE LEVELS DOUBLE?

Svante Ahremius took the first step toward getting a handle on how much an increase in carbon dioxide will affect average global temperatures. He figured that there would be a 5–6°C (7–9°F) increase for every doubling of carbon dioxide levels. Current estimates are lower partly owing to the impact of aerosols in the atmosphere that reflect incoming sunlight before it enters the atmosphere. Current IPCC estimates project that a doubling of carbon dioxide concentration from its present level would increase global average temperatures by 2–4.5°C (3.6–8.1°F). Different researchers have made different assumptions about the details of how the earth's climate system works, resulting in a range of estimates. The best estimate is that if atmospheric carbon dioxide concentration doubles, the global average temperature would rise by 3°C (5.4°F) and that it is very unlikely to produce an increase of less than 1.5°C (2.7°F).

TEMPERATURE INCREASES FOR VARIOUS SCENARIOS

A critical question on many people's minds is how quickly the temperature of the earth will rise. Models predict what the temperature will be for different levels of greenhouse gases. Various assumptions are included concerning how quickly greenhouse gases build up, where they are generated, and what is being done to limit emissions. Average global temperature ranges for various scenarios are shown in Figure 7-2. Each scenario results in different carbon dioxide levels. For instance, for the B1 scenario, the carbon dioxide level stabilizes at 540 ppm. Results are given for each of the following time periods: 2020s, 2050s, 2080s, and 2090s.

Regardless of which scenario is chosen, predicted temperatures are fairly tightly clustered in the forecast for the most recent time period, which is the 2020s. This means that we should not be lulled into a false sense of security if the near-term temperature increases are fairly modest. The real separation between scenarios begins to take effect toward the latter part of the twenty-first century, where more than 2°C (3.6°F) separates the most optimistic and pessimistic scenarios. Not surprisingly, the further out we get, the larger is the spread for a given scenario for a given time.

Higher greenhouse gas levels result in higher temperatures. The differences between high and low carbon dioxide levels will not be as pronounced in the next few decades as they will be toward the end of the twenty-first century.

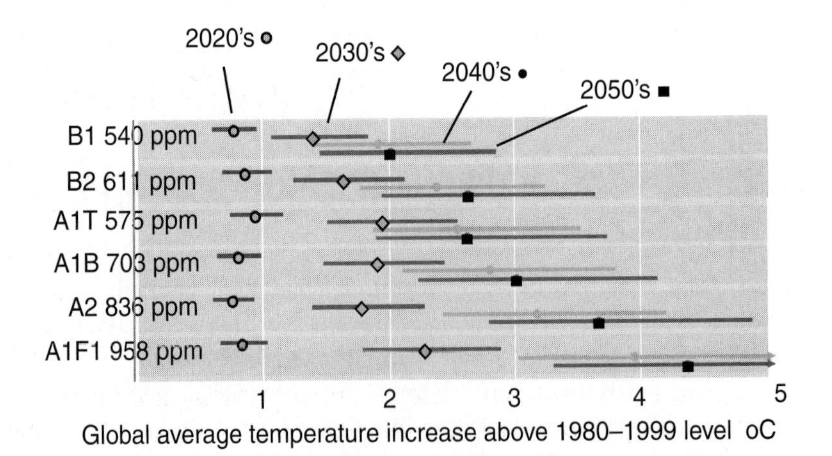

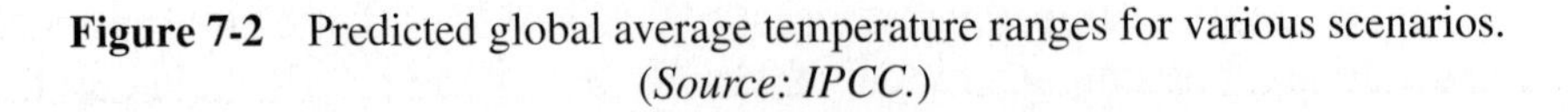
Figure 7-2 Predicted global average temperature ranges for various scenarios. (*Source: IPCC.*)

By the 2020s, the average global temperature is expected to increase 0.5–1.0°C (0.9–1.8°F) regardless of what carbon dioxide concentration is established. There are much bigger differences in temperature toward the latter part of the century for different concentrations. For instance, in the 2090s, the worst-case scenario (A1F1, stabilizing at 958 ppm carbon dioxide) results in an average global temperature that is 4°C (7.2°F) higher than the 1980–1999 level. The most optimistic scenario (B1, which has the earth stabilizing at 540 ppm) results in only a 1.5–2°C (2.7–3.6°F) temperature rise above the 1980–1999 level.

IMPACT AS TEMPERATURES RISE

Table 7-2 summarizes likely impacts for various increases in average global surface temperature during this century.

Climate models have been applied to the glaciers and ice sheets of Greenland, such as pictured in Figure 7-3. Results show that if carbon dioxide levels quadrupled from current levels, it would take 270 years for 20 percent of the ice to melt. After 1760 years, 80 percent of the ice would be gone. The progression of ice melting is shown in Figure 7-4.

HOW QUICKLY WILL TEMPERATURES INCREASE?

Let's assume for the moment that the world follows a "business as usual" path and continues burning fossil fuels that build up carbon dioxide in the atmosphere (this

Table 7-2 Likely Climate Impact of Various Increased Temperature Ranges

Range of Temperature Increase	Impact
+1°C (+1.8°F)	Decreasing water availability in the middle latitudes and dry areas Increased water in moist tropics and higher latitudes Increasing drought and risk of wild fires Increasing malnutrition Increased deaths from heat waves
+2°C (+3.6°F)	Risk of extinctions for 20–30 percent of known species Millions of people subjected to flood risk
+3°C (+5.4°F)	Widespread destruction of coral reefs About 30 percent of global coastal wetlands lost
+4°C (+7.2°F)	Global food production decreases Increased extinction risk Partial melting of Greenland ice sheet and West Antarctic sheet raising sea level by 4–6 m (13–20 ft)

Figure 7-3 Massive tongue of a Greenland glacier. (*Source: NASA.*)

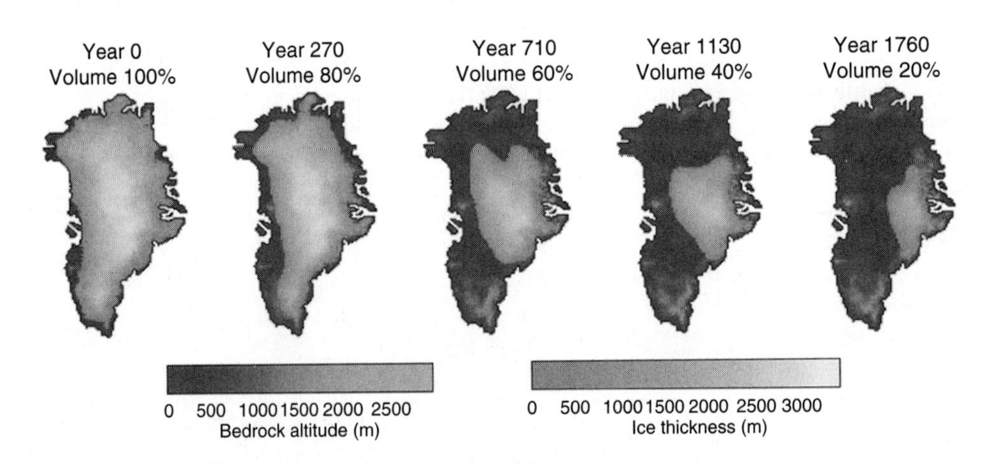

Figure 7-4 Model results showing massive melting of Greenland's ice sheet as a result of four times the current level of carbon dioxide. (*Source: IPCC.*)

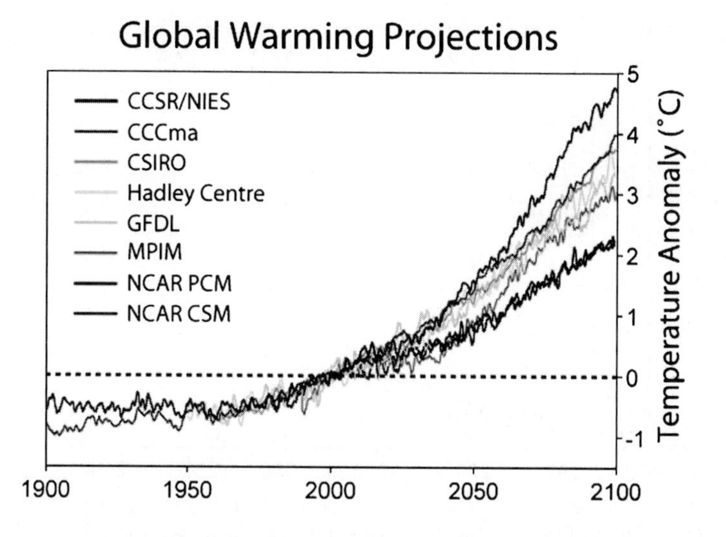

Figure 7-5 Climate model projections of the average global temperature increase above the 2000 baseline. (*Courtesy of R. Rohde, Global Warming Art, from IPCC data.*)

is the A2 scenario). By the end of the twenty-first century, temperatures are projected to increase by an average of 3.4°C (6.5°F) above what they were in 2000. Notice that the various models represented in Figure 7-5 all agree fairly well during the period of instrumental record. Two models have been applied to past climates to reconstruct temperatures derived from proxy measurements. The group of models predicts consistent near-term trends and reasonably similar long-term patterns. The further into the future the models forecast, the greater is the divergence based on differences in underlying assumptions and model details.

HOW RAPIDLY WILL THE SEA LEVEL RISE?

As the average global temperature increases, melting of land-based ice and thermal expansion cause the sea level to rise. Figure 7-6 shows how rapidly this is expected to occur.

By the end of the twenty-first century, continued emission of carbon dioxide from fossil fuel burning is expected to cause the sea level to increase. The various models forecast a rise of somewhere between 0.20 m (0.6 ft) and nearly 0.60 m (2.0 ft). A sea level rise less than 1 m by the end of the century will not cause the worst catastrophic coastal flooding. However, a 3°C (5.4°F) increase in temperature would continue to cause partial melting of the Greenland ice sheet and West Antarctic ice sheets and result in a 4- to 6-m (13–20 ft) increase in sea level within the next few thousand years. If global warming continues unchecked for centuries, complete melting of both the Greenland ice sheet and West Antarctic ice sheets is conceivable,

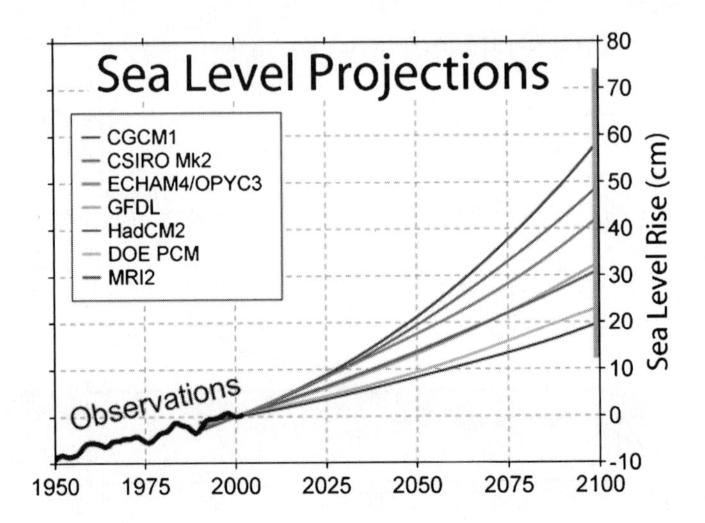

Figure 7-6 Climate model projections of sea level rise above the 1950–2000 baseline. (*Courtesy of R. Rohde, Global Warming Art, from IPCC AR4 data.*)

which could result in a sea level rise of 12 m (39 ft). The worse-case scenario (which is *not* a likely outcome of current climate conditions in the near future), in which there is complete melting of the all land-based ice, would raise sea level by 64 m (211 ft).

OUR WINDOW OF OPPORTUNITY TO ACT

Today, the carbon dioxide equivalent level (CO_2, eq) is about 430 ppm and is increasing by about 2 ppm per year. The *Stern Review on the Economics of Climate Change* (October 30, 2006) suggested that the worst impacts of global warming can be avoided if the carbon dioxide equivalent level is stabilized at between 450 and 550 ppm. This will require emissions to be decreased by 25 percent by 2050 and ultimately stabilized at no more than 80 percent below current levels. According to the report, if no action is taken, carbon dioxide levels could double by 2035. If this were to happen, the overall cost to the world would be at least 5 percent of the world's economic productivity. However, reduction of global warming would require an undertaking involving no more than around 1 percent of the world's gross domestic product. The conclusion is that "the benefits of strong and early action far outweigh the economic costs of not acting."

According to James Hansen[1] it takes about 30 years for 50 percent of the equilibrium temperature to be reached and about 250 years for 75 percent of the

[1]*James Hansen How Can We Avert Dangerous Climate Change? Paper, based on testimony to Select Committee on Energy Independence and Global Warming, U.S. House of Representatives, 26 April 2007, /www.columbia.edu/~jeh1/canweavert.pdf*

response and 1000 years for 90 percent of the response to be reached. This implies that climate changes that are already underway from past greenhouse gas emissions are in the pipeline. We have not felt the full brunt of their effects yet and should not be lulled into underestimating their impact based on the present state of the climate. Waiting to see what will happen can be dangerous because by the time the worst impacts of warming occur, it may be too late to respond. It also means that owing to this "climate inertia," it will take a long time to see the impact of any reductions of greenhouse gas emissions. Hansen proposed a limit of 450 ppm for carbon dioxide and that it is time now to "move promptly to the next phase of the industrial revolution."

POPULATIONS MOST AT RISK FROM SEA LEVEL RISE

The most vulnerable and least affluent countries will be affected the most and soonest despite the fact that they have contributed least to global warming. Looking at a satellite image of city lights gives an idea of how close the coastline populations are distributed. The image of North America shown in Figure 7-7 is typical of how population centers worldwide are clustered near the coastlines.

About 10 percent of the world's population lives less than 10 m (33 ft) above sea level (Figure 7-8), and the majority of the world's population lives within 10 km (6.2 miles) of sea level (Figure 7-9). A 5-m (16.4-ft) rise in sea level would cause serious flooding in many cities, including New York, London, Sydney, Vancouver, Mumbai, and Tokyo.

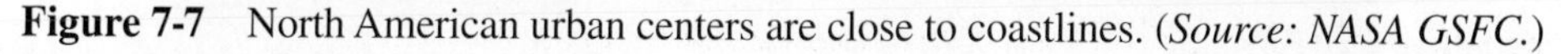

Figure 7-7 North American urban centers are close to coastlines. (*Source: NASA GSFC.*)

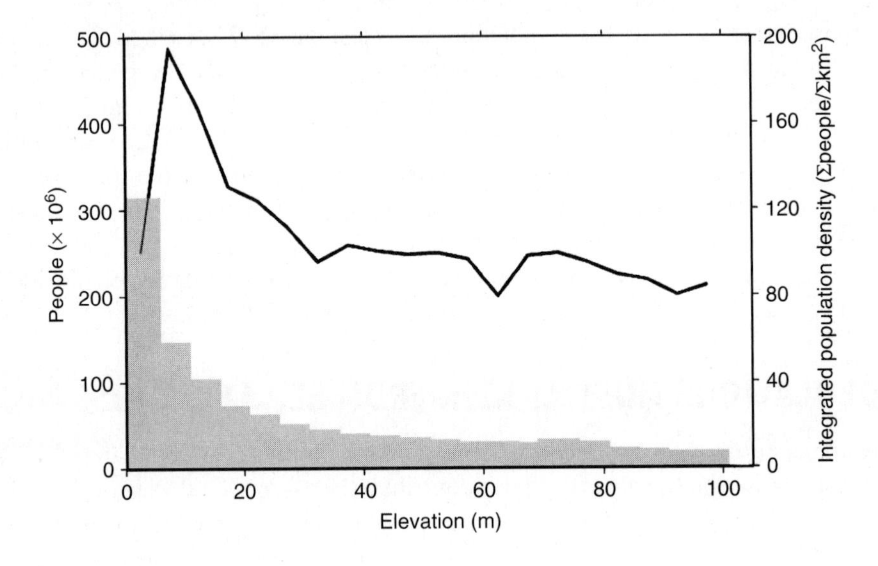

Figure 7-8 Much of the world's population is concentrated at elevations below 5–10 m (16.4–33 ft). (*Source: IPCC.*)

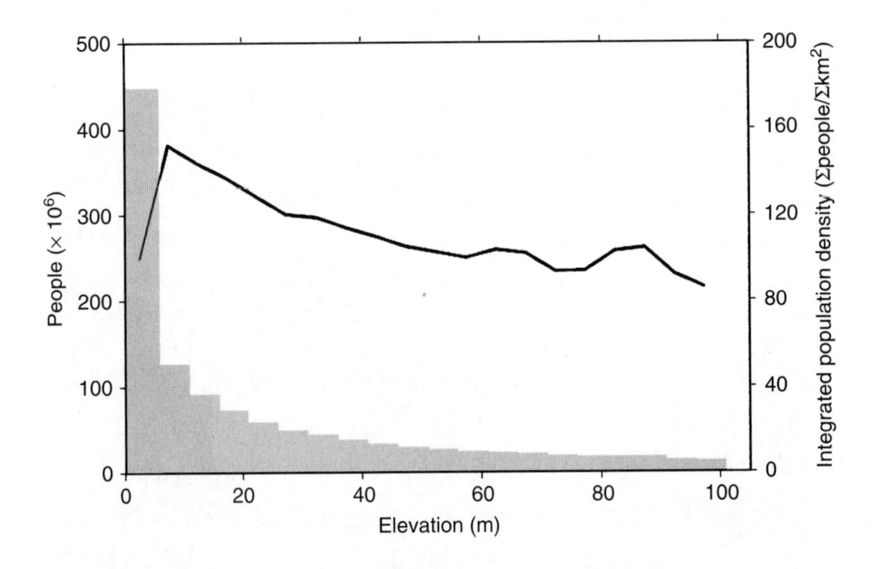

Figure 7-9 Much of the world's population is concentrated within 5 km (3.1 miles) of coastlines. (*Source: IPCC.*)

Impacts of Climate Change

The IPCC has indicated that climate changes currently taking place most likely will affect the following key areas.

WATER

Close to 70 percent of the earth's surface is covered with water, but only 1 percent of that water is drinkable. During this century, freshwater availability is expected to increase away from the equator and in some tropical regions by 10–40 percent. However, water will be less available where it is needed the most—in countries closer to the equator. Water supplies fed by melting glaciers and snow cover, after an initial surge, will begin to dwindle. One-sixth of the world's population currently relies on meltwater from major mountain ranges. As glaciers melt, some water supplies will have an abundance, if not an oversupply, of freshwater for a while. However, once the alpine snow cover melts, alternative sources of water will be needed in those areas. Water shortages are anticipated to be a major challenge to many parts of the world. Unless the world adapts to these changes, the water shortages could lead to economic disruption and conflict.

DROUGHTS AND FLOODS

A warmer planet is more prone to both droughts and floods. Since 1950, the number of heat waves around the world has increased. Drought is characterized by a combination of precipitation and temperature conditions. In Chapter 3 we saw a chart that showed a trend toward more droughts throughout the world based on the Palmer Drought Severity Index. There has been a drying trend since the mid-1950s over specific land areas, including Africa, southern Europe, southern Asia, Canada, and Alaska. Increased global temperatures lead to increased evaporation but also to an increased capacity of the atmosphere to hold water vapor. Climate data are not showing an overall global increase in precipitation. However, there is a statistical increase in the overall number of heavy precipitation events.

ECOSYSTEMS

The earth is, in many ways, forgiving of the abuses that people have subjected it to. However, we may be approaching a point where we will exceed the earth's ability to recover. If increases in average global temperatures start to increase by 1.5–2.5°C (2.7–4.5°F), many more plant and animal species face a greater risk of extinction. Some consequences of global warming no doubt will favor the survival of certain species, but the overall impact is to reduce the biodiversity of the earth. The decreasing pH level of the oceans (becoming more acidic) is likely to affect aquatic organisms that form shells, such as coral or other shellfish. This, in turn, will affect species that depend on the shell-forming organisms. A sea surface temperature increase of greater than 1°C (1.8°F) may be enough to initiate the destruction of coral reefs.

COASTAL AREAS

Many millions more people are projected to be exposed to flood conditions each year by the 2080s. About 1 billion people live within 25 m (82 ft) of today's sea level. These areas include many cities on the East Coast of the United States, nearly all of Bangladesh, and 250 million people in China. Relocation of all these people is unimaginable. Low-lying areas will be hardest hit and will be more vulnerable to storms that may become more intense and frequent. The low-lying river deltas of Asia and Africa pose some of the greatest risk, as do small islands near sea level. Figure 7-10 shows the locations of some of these threatened regions.

HEALTH

Effect of Climate

In some regions of the world, food productivity will improve. However, many areas will see a decrease in agricultural productivity because of the dual impacts of

Threatened Deltas

Relative vulnerability of coastal deltas as indicated by the indicative population potentially displaced by current sea-level trends to 2050 (Extreme ≥ 1 million; high 1 million to 50,000; medium 50,000 to 5,000)

Figure 7-10 River deltas around the world most vulnerable to flooding. (*Source: IPCC.*)

droughts and flooding. As a consequence, malnutrition may become more widespread. Heat waves, floods, storms, droughts, and fire are likely to have an impact. Fewer people are likely to die from cold exposure, but the negative health effects are more likely to outweigh the positive ones.

Disease

Waterborne diseases are responsible for 90 percent of the deaths from infectious disease worldwide. They may become more common on a planet that is becoming warmer, more polluted, and more heavily populated. Dengue fever is a viral disease characterized by fever, severe headache, and muscle and joint pain. A more dangerous form of this disease is dengue hemorrhagic fever, which broke out in Paraguay recently.

Dengue fever now affects 50–100 million people, mostly in tropical and subtropical regions of the world (*Science News*, March 15, 2005), and is spread by the *Aedes aegypti* mosquito. Climate change may increase the availability of standing water, which is where this mosquito breeds. With global warming, this mosquito also may migrate north. With warming conditions, there also may be some migration to higher altitudes, whose temperature range will become more conducive to mosquitoes. It is believed that this is occurring in Nairobi, Kenya. The regional distribution of other diseases, such as malaria, which is caused by a parasite spread by the female *Anopheles* mosquito, similarly may be affected by climate changes.

Displacement of mosquito-carried diseases beyond tropical and subtropical climates exposes a larger population to the disease. This may be offset in part by a better public health infrastructure to combat outbreaks of such diseases in those areas.

Regional Climate Changes

AFRICA

Water is expected to be a problem in Africa. By 2020, many millions of people will be affected by inadequate water supply. This will severely compromise the ability of the African people to grow food. As a result of climate changes, agricultural production throughout Africa may be cut in half by 2020.

ASIA

Throughout the Himalayas, melting glaciers during the next few decades are expected to produce local flooding and rock slides. After that, river flow will decrease, and

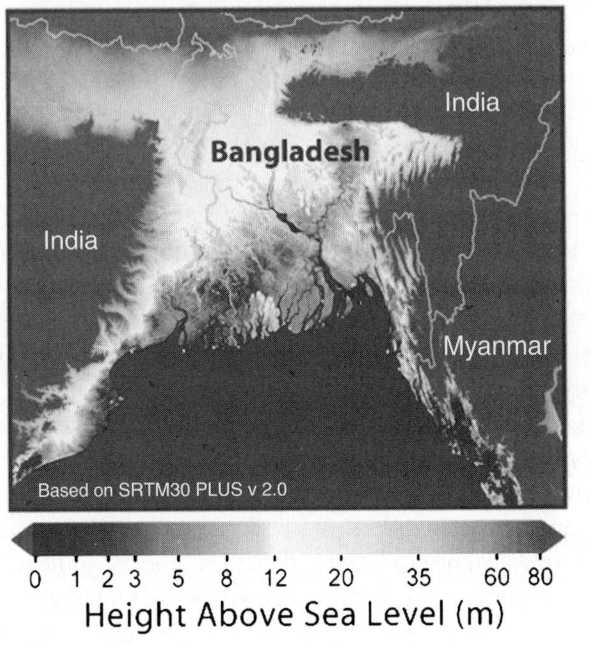

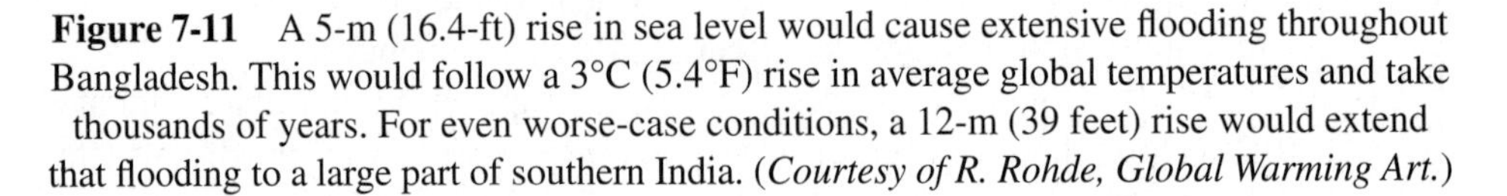

Figure 7-11 A 5-m (16.4-ft) rise in sea level would cause extensive flooding throughout Bangladesh. This would follow a 3°C (5.4°F) rise in average global temperatures and take thousands of years. For even worse-case conditions, a 12-m (39 feet) rise would extend that flooding to a large part of southern India. (*Courtesy of R. Rohde, Global Warming Art.*)

those who depend on flowing meltwater will see a decline in the availability of that resource. This could affect more than 1 billion people by 2050. Coastal flooding, especially during storms, will become an increasing concern. (Figure 7-11 shows how this could play out after many centuries of increased atmospheric temperatures.) Agriculture actually may pick up in the east and southeastern parts of Asia but is expected to fall in the central and southern parts of the continent. Combined with rapid population growth, the risk of hunger in that part of the world may become severe by the middle of the twenty-first century.

AUSTRALIA AND NEW ZEALAND

Decreased precipitation and increased evaporation will increase the risk of drought in this part of the world. The tendency toward increasing concentration of populations near the coast increases the impact of floods and coastal storms. Droughts and wildfires are anticipated in many areas, although some parts of New Zealand many benefit initially from reduced areas prone to frost.

EUROPE

The heat waves and flooding that have ravaged parts of Europe already are expected to continue. Glacial melting will affect water supplies. Short-term benefits may include a reduced demand for heating, increased crop yields, and increased forest growth.

LATIN AMERICA

A significant shift in climate from tropical forest to savannah in the regions east of the Amazon may occur by midcentury. Drier conditions are expected to lead to desertification of some agricultural land. Some crops will no longer be viable, and livestock production may need to be relocated. Some improvements in crop production in temperate regions is possible.

NORTH AMERICA

Decreased snowpack in mountainous areas will contribute to more winter flooding and less flow in summer. There is currently not enough water in the western part of North America for the many intended users, and global warming will make this situation even more difficult. In areas that receive less precipitation, there is an increased risk of brush and forest fires. Farm production may improve by 5–20 percent in some areas where growing seasons are lengthened. Cities may feel a more severe impact from the heat island effect coupled with global warming. A long-term projection of the impact of coastal flooding is shown in Figure 7-12.

POLAR REGIONS

Melting of ice sheets is expected to lead to a loss of habitat for a number of organisms, including migratory birds, polar animals, and their predators. In the Arctic, the impact on human communities is expected to be mixed. Heating requirements will be reduced, and northern sea routes will become navigable. Traditional ways of life will be threatened, and the indigenous polar populations will be confronted with the need to relocate or adapt to the changing conditions.

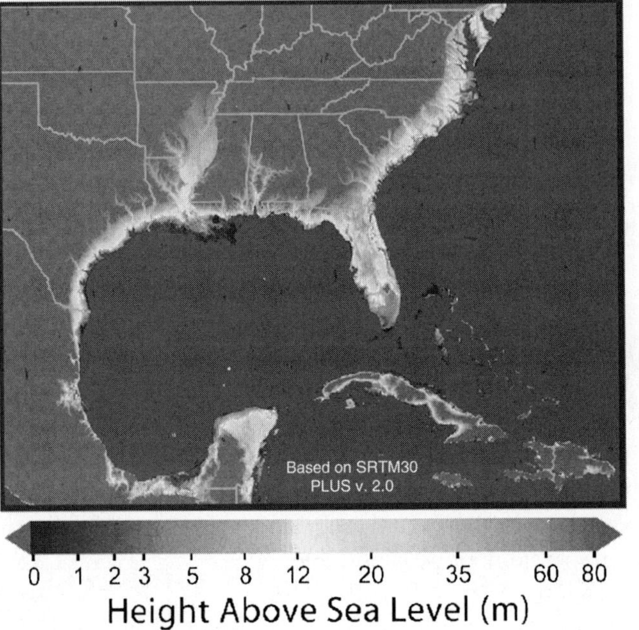

Figure 7-12 An increase in sea level of 5 m (16.4 ft) would inundate much of the east coast of the United States including southern Florida, the Gulf Coast, and many Caribbean Islands. This would follow a 3°C (5.4°F) rise in average global temperature following thousands of years of warming. (*Courtesy of R. Rohde, Global Warming Art.*)

SMALL ISLANDS

Erosion of beaches and loss of coral reef in nearby coastal areas will affect people living on small islands throughout the Pacific and Caribbean. Rising sea levels will pose a greater hazard during storms. Less potable water may be available for direct consumption or agriculture.

Back to the Future—Will the Past Repeat Itself?

The geologic history of the earth has been one of alternating warm and cool periods. It may be helpful to look back and understand better what changes have occurred in the past and what insights they may provide for us today. We are currently in a warm phase, but it is not as warm as the extremes of previous interglacial warm periods.

Based on the orbital forces that have driven this cycle, we are *very slowly* headed for another ice age. According to the IPCC, it is "virtually certain" that natural processes will cause the earth to cool. However, this will not happen for at least 30,000 years. Before that, the earth can be expected to experience a period of global warming caused by the human-enhanced greenhouse effect.

Past warming periods have been accompanied by increases of greenhouse gases such as carbon dioxide and methane. This, however, is seen by climatologists as more an effect rather than a cause of the warming that brought the earth out of past ice ages. Ice-core measurements show that temperatures in Antarctica started to rise centuries *before* the carbon dioxide levels started to increase. Some observers have cited this as reason to dispute the entire concept of greenhouse gas–induced global warming.

Clearly, the elevated carbon dioxide levels that accompanied the earth's recovery from past ice ages were not the result of fossil fuel combustion. Some scientists have considered the possibility that elevated carbon dioxide and methane levels came from volcanic activity.

According to the IPCC, it is very unlikely that carbon dioxide triggered the end of the ice ages, but the elevated carbon dioxide did help the earth to warm up. The greenhouse gases amplified the warming effects brought by cyclic changes in the earths orbit by providing positive feedback.

The current levels of carbon dioxide and methane in the air today have never been experienced before, leaving the earth in uncharted waters. The levels of the major greenhouse gases—carbon dioxide and methane—are far greater than they have ever been during the past 650,000 years. The rate of increase is also greater than it has ever been at any time during the past 16,000 years.

In the past, when the carbon dioxide and methane levels were at higher levels, the earth was warmer owing to the amplifying effect of the greenhouse gases. Since the last glacial maximum (ice age), the average global temperature likely has increased by 4–7°C (7.2–12.6°F). During that interglacial period, sea levels were about 4–6 m (13–20 feet) higher than they are at present. Greenhouse gas concentration changes in the past occurred much more slowly than at present. Today, greenhouse gases are changing at a rate that is about 10 times faster.

The effect of positive feedback can accelerate the onset of a climate change. When the tongues of glaciers are released into the sea, the movement of the ice mass increases. Surface melting on ice layers increases absorption and results in faster melting. These possibilities lead to concerns about the likelihood of abrupt climate changes.

One abrupt change happened about 14,500 years ago. The earth's climate at the time was in the process of changing from a cold glacial period to a warmer interglacial period when temperatures in North America rapidly returned to near-ice-age conditions. This period is known as the *Younger Dryas* (named after a flower that grew in cold conditions). The Younger Dryas Period also ended abruptly (Figure 7-13).

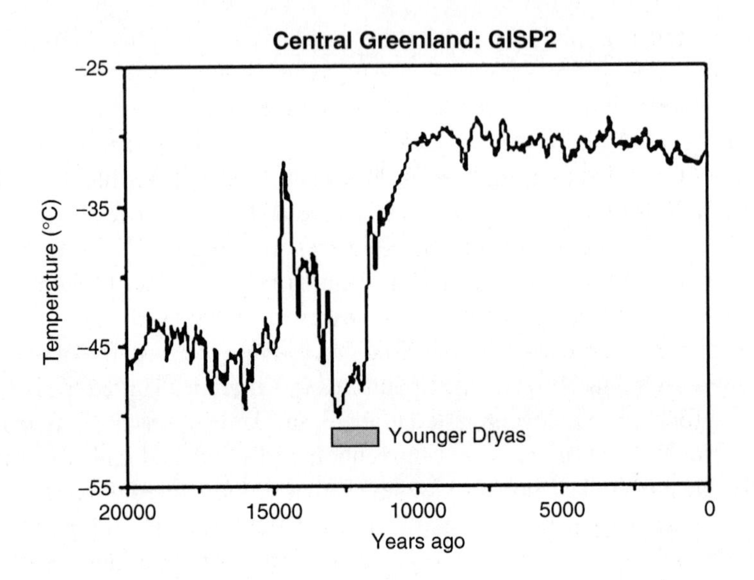

Figure 7-13 The abrupt cooling and warming in Greenland that occurred during the Younger Dryas Period. Temperature measurements are based on GISP2 ice-core data and illustrate an historical example of abrupt climate change. (*Courtesy: NOAA; Credit: Cuffey Cloy and Alley.*)

Scientists think that this may have been the result of Greenland ice melting, adding freshwater to the North Atlantic. The decrease in salinity may have put a damper on the thermohaline circulation (THC) that drove the Gulf stream. With less heat being distributed by this massive thermal conveyor belt, the abrupt change may have been triggered. Scientists are asking whether other similar abrupt changes may be initiated by global warming magnified by positive feedbacks.

FILM REVIEW: THE DAY AFTER *THE DAY AFTER TOMORROW*—GOOD LESSONS FROM BAD SCIENCE

The Plot

Jack Hall (Dennis Quaid), a paleoclimatologist, predicts that the earth is about to enter an ice age triggered by global warming. This prediction is fulfilled as a sequence of events triggers abrupt climate change, creating a global superstorm and weather disasters. In the span of just a few days, tornados roll through Los Angeles, massive hail falls on Tokyo, and blizzards tear into New York.

The movie: As a result of sea ice melting and a storm surge, the water surrounding the Statue of Liberty is 150–215 ft (45.5–65.5 m) above the present sea level.

The science: The maximum amount of potential sea level rise from all ice caps and glaciers melting, as discussed in Chapter 3, is 62 m (211 ft). If this melting is the result of greater solar heat retention caused by greenhouse gases, the phase change from ice to liquid water would take many years—not hours, as depicted in the movie.

The movie: Rapid release of freshwater caused an abrupt shutdown of thermohaline circulation, disrupting the Gulf stream. This instantly plunged the North Atlantic into an ice age.

The science: The IPCC predicts that ice melting from Greenland could slow the thermohaline circulation. However, a complete shutdown is not likely or necessarily a done deal for at least hundreds or thousands of years. Without the Gulf stream, average temperatures would cool in Europe and North America by perhaps 2.8°C (5°F). "Abrupt" climate change (such as onset of the Younger Dryas Period) has occurred on a time scale of years rather than days, as was depicted in this movie. In the film, thermohaline circulation disruption occurred too rapidly, affected climate too abruptly, and had consequences that were too severe. Otherwise, granting the filmmakers some license for exaggeration, this aspect of the film has some basis in reality, but the portrayal is influenced more by drama than by science.

The movie: The superstorm brings cold upper atmospheric air down to the surface, causing severe and rapid freezing.

The science: According to the ideal gas law, descending air would warm up (as its pressure increases) rather than cool. The thinner air in the upper troposphere also would have a small heat capacity and as a result little ability to flash freeze anything in its path.

Results of Climate Models—Levels of Certainty

PREDICTIONS

- *Global average temperature.* The global average temperature is expected to rise by 0.64–0.69°C (1.15–1.24°F) by the years 2011–2030 compared with the years 1980–1999. This result is the consensus of three different models used by IPCC investigators.
- *Areas most affected.* The greatest temperature increases will occur at high northern latitudes (such as Alaska and the Arctic regions) and over land areas.

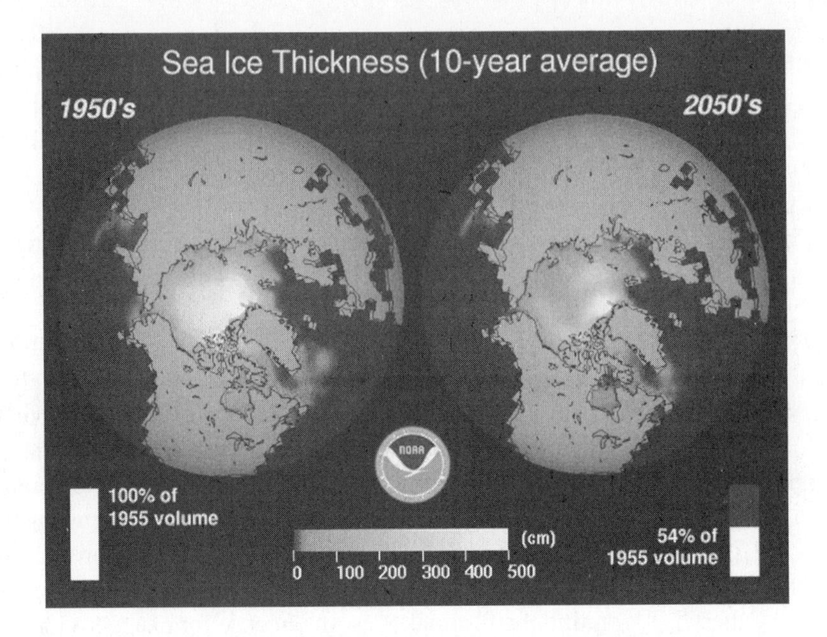

Figure 7-14 Ten-year average sea ice thickness at the north pole is expected to decline by 54 percent by the 2050s. (*Source: NOAA.*)

The Arctic Ocean is *likely* to be free of ice during the summer months within roughly five decades. The Arctic sea ice overall is expected to lose 54 percent of its volume by the 2050s. (See the model projections in Figure 7-14).

- *Ocean circulation.* Some investigators have expressed concern that the ocean circulation patterns that keep northern European countries in a more temperate range may be disrupted by the effects of global warming. The IPCC predicts that the massive Atlantic circulation flow called the *thermohaline circulation* (and also referred to as the *meridional overturning circulation,* or MOC) "is *very likely* to slow down by 2100. However, according to the IPCC, it is *very unlikely* that the MOC will undergo a large abrupt transition during the course of the twenty-first century"
- *Ice sheet melting.* At least partial melting of the Greenland ice sheet and the West Antarctic ice sheet is expected if there is a global temperature increase in the range of 1–4°C (1.8–7.2°F) compared with 1990–2000. Because of the many uncertainties associated with this outcome, the IPCC places what it calls "medium confidence" in this outcome, in contrast with other predictions that are known with a greater degree of certainty. This melting may take centuries or even thousands of years.

- *Weather extremes.* It is considered very likely that heat waves and heavy precipitation events will continue and become more frequent. However, observations of small-scale severe weather such as tornadoes are local and too scattered to draw conclusions from. To some extent, greater public awareness and better tracking of these events may inflate records.
- *Drought.* Climate data show a large drying trend over many areas of North America, southern Europe and Asia, northern Africa, Canada, and Alaska. An index used to measure the occurrence of drought conditions around the world that combines the local effects of both water shortage and elevated temperature is shown in Figure 7-6.
- *Antarctica ice sheet.* Models suggest that the Antarctic ice sheet as a whole will remain too cold to melt and will gain in mass as a result of increased snowfall.

HOW ACCURATE ARE THE MODELS?

Much has been accomplished in the science of climatology in recent years. Some aspects of the earth's climate are known with a high degree of confidence. As might be reasonably expected, there are areas where the jury is still out and further work remains to be done. Much of this centers around gaining a more precise understanding of past climate changes. In many ways, the scientists and modelers seem to be attempting the technical equivalent of herding cats. But we are far from clueless.

A standard scientific practice is to determine the accuracy of a particular set of measurements and consequently the conclusions based on those measurements. Uncertainty can be either in (1) determining a particular measured value such as temperature, sea level, or mass of an ice sheet or (2) the cause-and-effect relationships between the variables, such as how much new carbon dioxide will the oceans absorb, how much warming will result from a given amount of greenhouse gas, or how much warmer will the North Atlantic need to be to completely cause the Greenland ice sheet to melt.

Since some outcomes have a much greater degree of certainty than others, the IPCC has established a consistent terminology to use in describing climate predictions. For instance, the following statements are examples of the results of models used to simulate the effects on climate:

- It is *very likely* that the average northern hemisphere temperatures during the second half of the twentieth century were warmer than any other 50-year period in the last 500 years.
- It is *very likely* that human (anthropogenic) greenhouse gas increases caused most of the observed increases in global average temperatures since the mid-twentieth century.

- It is *virtually certain* that human-generated aerosols have a cooling effect on northern hemisphere air temperatures.
- It is *extremely likely* that human activities have exerted a warming effect.
- It is *extremely unlikely* that the earth would naturally enter another ice age for at least 30,000 years.
- It is *likely* that there have been increases in the number of heavy precipitation events.
- There is *high confidence* that the rate of sea level rise has accelerated between the mid-nineteenth and twentieth centuries.

Each of the italicized terms indicates a level of certainty that has a specific statistical probability associated with it, as shown in Table 7-3.

POSSIBLE AND PROBABLE OUTCOMES

One of the most likely outcomes of climate change with the greatest degree of certainty attached to it is an increased incidence of warmer days and nights and a reduced incidence of cold days and nights. Figure 7-15 shows the trend for warm nights (defined as above the 90th percentile range). The vertical scale shows days per 10 years.

Table 7-4 lists the likelihood of various trends, and Table 7-5 summarizes some possible impacts of climate changes anticipated during the twenty-first century according to the IPCC.

Table 7-3 IPCC Statistical Confidence Ranges

Term	Probability
Virtually certain	>99 percent
Extremely likely	>95 percent
Very likely	>90 percent
Likely	>66 percent
More likely than not	>50 percent
About as likely as not	31–66 percent
Unlikely	<33 percent
Very unlikely	<10 percent
Extremely unlikely	<5 percent
Exceptionally unlikely	<1 percent

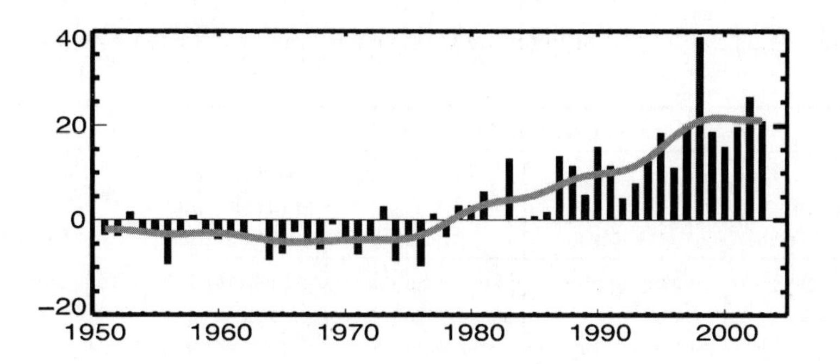

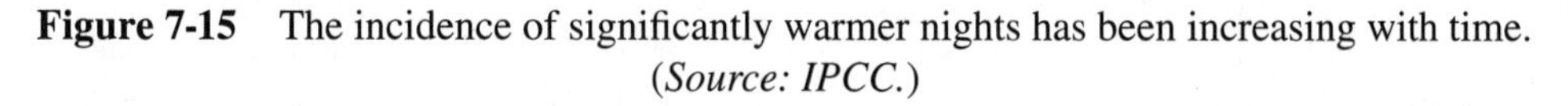
Figure 7-15 The incidence of significantly warmer nights has been increasing with time. (*Source: IPCC.*)

BASIC UNCERTAINTIES—WHERE THE JURY IS STILL OUT

In science, it is usually valuable to know what you do not know. Defining the limitations of knowledge clarifies areas for follow-up effort and separates logical

Table 7-4 Likelihood of Various Trends

Trend	**Likelihood that Trend Occurred in the Late Twentieth Century**	**Likelihood of a Human Contribution to Observed Trend**	**Likelihood of Future Trends in the Twenty-First Century**
Warmer and fewer cold days and nights (over land areas)	Very likely	Likely	Virtually certain
Warmer and more frequent hot days (over land areas)	Very likely	Likely	Virtually certain
Warm spells/heat waves becoming more frequent	Likely	More likely than not	Very likely
Heavy precipitation events becoming more frequent	Likely	More likely than not	Very likely
More areas affected by droughts	Likely (in many places since 1970)	More likely than not	Likely
Intense tropical cyclone activity increases	Likely (in some places since 1970)	More likely than not	Likely
Increased4 incidence of extreme high sea level	Likely	More likely than not	Likely

Table 7-5 Potential Impact of Potential Climate Changes in Affected Areas

Likelihood	Areas Likely to Be Affected by Climate Change				
	Overall Climate Trend	**Water**	**Health**	**Agriculture, Forests, and Ecosystems**	**People and Industry**
Virtually certain	Warmer days and nights	Water supplies that rely on snow melting are affected	Fewer human deaths from cold exposure	More crops in cold areas Fewer crops in warm areas Insect abundance	Reduced heating requirements Increased cooling demand Fewer disruptions from winter storms
Very likely	More frequent warm spells/ heat waves	Increased demand for water Algal blooms, etc.	Increased risk of heat related deaths (especially in vulnerable populations).	Reduced crop yields in warm areas More wildfires	Impacts on elderly, very young, and poor
Very likely	More frequent heavy precipitation	Adverse effects on surface and groundwater Contamination of water supply Possible relief of scarcity	Increased risk of death, injury, and disease	Crop damage Soil erosion Loss of farmland	Disruptions owing to flooding Infrastructure damage Loss of property
Likely	More areas exposed to drought	Water shortages	Food and water shortages Risk of malnutrition Increased risk of disease	Land degradation Lower crop yields Livestock deaths Increased risk of wildfire	Water shortages Reduced hydroelectric power Possible population relocations
Likely	More intense tropical cyclone activity	Water supply disruptions from power outages	Increased death, injury, and disease	Crop damage Tree damage Damage to coral reefs	Disruption from flood and high wind Loss of insurance Possible population relocations
Likely	Increased incidence of high sea levels	Saltwater intrusion contaminating freshwater	Increased risk of death and injury	Increased salt levels in irrigation water	Higher cost of coastal protection Possible population relocations

conclusions from speculation. As much as climatology has evolved in recent years, there remain the following areas of uncertainty:

1. The effect of solar intensity variability for the past few centuries is not clear because instruments to precisely make this measurement were not available at the time. Insight into the effects of changing solar intensity on past climates can only be derived from indirect or proxy data.
2. The detailed interaction between aerosols and cloud properties is not well understood.
3. Causes of the recent reduction of methane in the atmosphere are subject to various theories but are not clearly understood.
4. Details of stratospheric water vapor (such as contributed by airplane contrails) are not well quantified.
5. Some radiosonde balloon data may be unreliable, especially in tropical areas.
6. Information on hurricane frequency and intensity is limited to recently acquired satellite data. This makes it difficult for observers to determine with confidence whether there is a trend toward more severe weather.
7. Similarly, there is insufficient evidence to determine whether tornados and other severe weather are intensifying.
8. Prior to 1960, there was no global measurement of overall snow cover.
9. There is not enough information to draw a conclusion about trends in the thickness of Antarctica sea ice.
10. The global average sea level rise measured between 1961 and 2003 is larger than can be fully accounted for by thermal expansion and land-based ice melting.
11. Mechanisms for past abrupt climate change (such as the Younger Dryas Period) are not well understood. Thresholds for when abrupt changes may occur are also not nailed down with precision.
12. Historical (paleoclimate) records are available for the northern hemisphere, but fewer records exist for the southern hemisphere.
13. Factors affecting temperature changes are much better understood than those influencing precipitation.
14. Processes taking place in the ocean depths that influence climate are more difficult to model.
15. Models suggest that the thermohaline circulation (or meridional overturning circulation) will not be disrupted by the end of the century.

Greater uncertainty exists in predicting how greenhouse gas emissions will affect the MOC after 2100 and whether there is a threshold that could trigger abrupt changes.

16. The ENSO is only partially understood and is not modeled the same way by all scientists.
17. There is uncertainty about the amount of sustained global warming that would lead to complete melting of the Greenland ice sheet.
18. There is very limited correlation between climate variables and the incidence of extreme events.
19. During past ice ages, the carbon dioxide level in the atmosphere dropped. Although various possible explanations have been proposed, the precise mechanism causing this has not been determined.
20. Perhaps the greatest uncertainty is not knowing what actions the inhabitants of the earth will take in response to the climatic changes that are known with higher levels of certainty.

Key Ideas

- Climate sensitivity describes how much global average temperature increases when the carbon dioxide level in the atmosphere doubles.
- If the carbon dioxide level in the atmosphere doubles, the global average temperature is expected to increase by 3°C (5.4°F) within a range of 2–4.5°C (3.6–8.1°F).
- Climate models predict climate changes based on quantitative relationships between the variables and feedback.
- Feedback occurs when the results of one process affect the outcome of a second process, which, in turn, affects the outcome of the first process.
- Positive feedback causes an outcome to be more intense than it otherwise would have been.
- Negative feedback causes an outcome to be less intense than it otherwise would have been.
- Current global warming is expected to cause disruptions to water supplies (initial excesses followed by shortages), intensified droughts and floods, and more severe weather.

- The global average temperature is expected to rise by 0.64–0.69°C (1.15–1.24°F) by the years 2011–2030 compared with the years 1980–1999.
- Ocean circulation processes (thermohaline circulation) are expected to slow but not be disrupted entirely.
- The Arctic Ocean may have no ice at the end of the summer of 2060.
- The Antarctic ice sheet as a whole will remain too cold to melt entirely and will gain in mass as a result of increased snowfall.

Review Questions

1. Which of the following is an example of positive feedback?
   (a) Increased evaporation produces clouds that reflect more sunlight.
   (b) Increasing carbon dioxide in the air promotes plant growth, which removes additional carbon dioxide from the air.
   (c) The darker surface of melting of snow on glaciers and ice caps reflects less sunlight.
   (d) The warmer the atmosphere, the more effectively it radiates infrared radiation to space.
2. Which of the following is an example of negative feedback?
   (a) Increased temperature increases evaporation, which produces greater absorption of sunlight by water vapor in the air.
   (b) Increased temperature results in drought, which causes plants to die and decay. More carbon dioxide is released when the plants decay.
   (c) Higher ocean temperatures reduce the solubility of carbon dioxide in the oceans.
   (d) Increasing carbon dioxide in the air promotes plant growth, which removes additional carbon dioxide from the air.
3. What is the *most likely* outcome if greenhouse gas emissions are held constant at today's levels?
   (a) Global temperature would continue to increase.
   (b) Global temperatures would begin to decrease immediately.
   (c) Global temperatures would begin to decrease after about 2 years.
   (d) global temperatures would be held constant.

4. Suppose that the carbon dioxide concentration in the atmosphere doubles from its present level. How much warmer would the average global temperature most likely be?
   (a) 0.5°C (0.9°F)
   (b) 1.0°C (1.8°F)
   (c) 3.0°C (5.4°F)
   (d) 5.5°C (9.9°F)
5. How is global warming expected to affect the thermohaline circulation (meridional overturning circulation, or MOC) in the ocean in the near future?
   (a) Complete disruption, causing a much colder European climate
   (b) Slowing, but circulation pattern continues and remains intact
   (c) No effect at all
   (d) Intensification of circulation pattern, causing warming of Europe
6. What role is carbon dioxide thought to have played in ending past ice ages?
   (a) Increased levels of carbon dioxide caused the earth to begin to warm up.
   (b) Carbon dioxide increased in the atmosphere as a result of the warming trend and amplified the warming effect.
   (c) Carbon dioxide did not play any role in past ice ages.
   (d) Carbon dioxide slowed withdrawal from the ice ages by imposing a negative-feedback effect.
7. Which of the following is an area that does *not* have a great deal of scientific understanding at present?
   (a) The temperature increase for a given greenhouse gas concentration
   (b) How much average global air temperature has increased in the last century
   (c) The interaction of aerosols and cloud formation
   (d) The net impact of aerosol on the atmosphere
8. What is a climate commitment?
   (a) How many hours climate scientists work
   (b) How much warmer the atmosphere will get based on past emissions of greenhouse gases
   (c) The impact on temperature from natural causes
   (d) What different countries decide to do after reaching a stabilization level of emissions

9. Which of the following outcome has the *greatest probability* of occurring?
   (a) Melting of Antarctic permanent ice
   (b) Significantly higher sea levels
   (c) Spread of dengue fever throughout North America
   (d) Fewer cold days and nights (over land areas)
10. What temperature increase above present levels is expected to initiate significant melting of the Greenland and Antarctic ice sheets resulting in a sea level increase about 4–6 m (13–20 ft).
   (a) +0.5°C (+0.9°F)
   (b) +1°C (+1.8°F)
   (c) +4°C (+7.2°F)
   (d) +8°C (+14.4°F)

CHAPTER 8

Resetting the Earth's Thermostat—Solutions

The inhabitants of the earth are now at a crossroads. The choices—whether they are made deliberately or by default—are where and when we stabilize the level of greenhouse gases and how we respond to the possible consequences of not having done that soon enough. This chapter will explore actions that can help to mitigate the changes that have been set in motion. We also will deal with the options that are available to adapt to the climate changes that will occur regardless of what actions we take.

This chapter will focus primarily on the two major sources of greenhouse gases: coal-burning electricity-generating plants and petroleum-burning vehicles. Together these contribute 80 percent of the heat-absorbing gases we are loading into the atmosphere. Solutions are available, but most assuredly they will require a departure from doing business as usual.

Electrical Power: The Problem with Coal

WHAT'S NOT TO LIKE ABOUT COAL

Coal is cheap and plentiful. It fueled the industrial revolution in Europe and then in the United States. The emerging countries still see coal as providing the same opportunity for growth and prosperity that it afforded the developed nations. China is a leading producer of coal worldwide, meeting almost one-third of the world's demand in 2005, followed by the United States and India (British Geological Survey). The United States (as well as other countries) has an abundant supply of coal, enough to last at least a century, providing a major component of America's long-sought energy independence.

Coal plants work by burning coal, which, in turn, boils water. That water turns to steam, which turns a turbine, which turns a generator, which produces the electricity. Figure 8-1 shows a typical coal-fired power plant.

There is no doubt that coal will continue to play a major role in producing the world's electricity for years to come. Electrical power companies in the United States are expected to add 280–500 megawatt power plants by 2030. By that time, the newly installed electrical power plants throughout the world may very well have added as much carbon dioxide to the atmosphere as was added during the entire

Figure 8-1 Most electricity in the world is generated by coal-fired electrical power plants. (*Photo by Ally Silver.*)

industrial revolution. China is on pace to construct the equivalent of one large coal-fired electricity-generating plant each week. It is inevitable that coal will continue to be used to provide the lion's share of electricity around the world in the near future. To come to terms with global warming, it is necessary to squarely confront the challenge faced by emissions from coal-fired electrical power plants.

CARBON CAPTURE AND STORAGE—SEQUESTRATION

There is no way to prevent coal from producing carbon dioxide when it burns. However, it *is* possible to prevent the carbon dioxide that is generated from being released into the atmosphere. In this scenario, carbon dioxide is separated from the exhaust gases, moved, and then stored. The process is called *capture and storage* (CSS) or *geologic carbon sequestration* using equipment such as that shown in Figure 8-2. Possible storage methods under consideration include injection in stable underground geologic formations, dissolution in the ocean depths, or binding in solid form as chemical carbonates.

The idea of removing products of combustion from a power plant smokestack is not entirely new. Since the Clean Air Act of 1970, power plant operators have installed equipment that removes particulates and the oxides of sulfur and nitrogen before the waste gases are released into the air. In a similar manner, carbon dioxide also can be removed from the output of the smokestack.

Similar capture technology is used extensively throughout the world in the manufacture of chemicals such as fertilizer and in the purification of natural gas. Storage in underground reservoirs is the most mature and probably most likely sequestration approach. A similar technique is used by the petroleum industry.

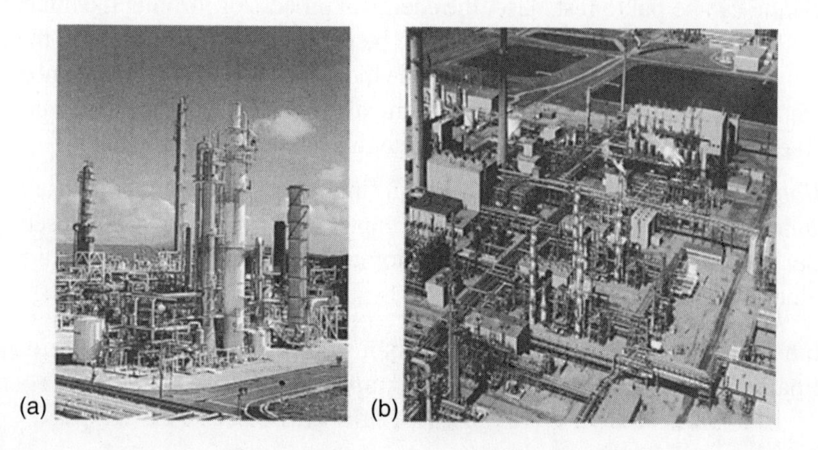

Figure 8-2 Carbon dioxide sequestration site. Similar facilities could separate carbon dioxide emissions and transport them for permanent storage. (*Source: IPCC.*)

Carbon dioxide is injected underground to help force the last drops of oil out of fields that are nearly empty.

The Intergovernmental Panel on Climate Change (IPCC) estimates that there are sufficient locations throughout the world to capture all the carbon dioxide that is likely to be generated from all fossil fuel–consuming plants operating through the twenty-first century. The closer the reservoir is to the power plant, the lower is the cost of capturing and storing its carbon dioxide emissions. A recent study by Hawkings[1] et al., 2006 estimates that capture and storage would add an extra 1.6 cents to an average cost of 4.7 cents/kWh for coal-generated electricity.

This would make coal-generated electrical power that is free of carbon dioxide approximately 25 percent more expensive. Carbon captured from coal-burning plants can be transported economically to sites that require carbon dioxide to enhance oil recovery. If this occurs, much, if not all, of this extra cost can be offset.

Approaches to sequestration include the following:

Geologic sequestration. Captured carbon dioxide is stored underground in sites such as depleted oil and gas fields, coal seams, and brine fields. Injection of carbon dioxide into coal seams has the added potential benefit of producing methane, which could be extracted as a fuel. Geologic sequestration is effective only if the containment is secure over the long term. Any leak of carbon dioxide over time defeats the purpose of storing it in the first place.

Chemical sequestration. Carbon removed from smokestacks could be secured by forming stable minerals such as calcite ($CaCO_3$) or magnesite ($MgCO_3$). This parallels the weathering process that results in the formation of natural minerals and may be less likely to trigger unforeseen ecologic consequences. The minerals could be expected to be stable for millions of years, so concerns about leaking can be put to rest. Also, the chemical process of forming the minerals is exothermic, so the heat generated could be put to use and contribute to the overall energy efficiency of the operation. Since the ash leftover from coal combustion contains calcium and magnesium (chemical constituents of the stable minerals), experiments are underway to incorporate this into sequestration tests.

Biologic sequestration. This involves fixing the carbon dioxide in biomass such as algae (described later in this chapter). The biomass can be used as a fuel or a food source. As with other biomass solutions, the carbon ultimately is released to the atmosphere.

Some new power plants, however, are being configured as "sequestration ready," anticipating the possible move toward its implementation.

---

[1]Hawkings, Lashof, and Williams. *What To Do About Coal?, Scientific American*, September 2006, p. 69.

BURNING COAL MORE EFFICIENTLY

Standard coal-fired electricity-generating plants burn the coal in air, which consists of nearly 80 percent nitrogen gas. The heat generated produces steam, which turns a turbine, which turns a generator. Removal of the carbon dioxide from the gas stream is difficult because of the large amount of nitrogen gas that is also coming out of the smokestack.

A more advanced approach to burning coal is called *integrated gasification combined cycle* (IGCC). In this approach, the coal is first treated with steam. The coal is partially oxidized to produce carbon monoxide and hydrogen gas. This combination is known as *synthesis gas*, or *syngas*. The syngas then is burned and further treated, resulting in a much higher concentration of carbon dioxide. The carbon dioxide then can be removed much more easily at lower overall cost.

Capture and storage processes will compromise the overall efficiency of the process of converting coal to electrical energy. As a result, conventional coal-fired plants may need to consume 30 percent more coal, and IGCC plants may require 20 percent more coal. The benefit will be a reduction of carbon dioxide emissions in a way that actually could benefit rather than hurt the coal industry.

To accomplish capture and storage of carbon dioxide from coal-generated plants, electric utility bills may need to be more than 30 percent higher than at present.

USING LESS COAL—CONSERVATION OF ELECTRICAL ENERGY

Coal burning is a major source of carbon dioxide. It is often said that the least expensive source of energy is conservation. The less electricity we use, the less coal we need to burn to provide it.

Electrical Energy—How Much Do We Need?

The United States has the highest use of energy per person of any other country in the world. The typical home in the United States has an average power consumption of about 1 kW (1000 W) of electricity. This is the equivalent of having ten 100-W light bulbs burning all the time. This creates a typical annual energy requirement of 8000–10,000 kWh for each home. There are many opportunities to reduce this requirement.

Table 8-1 illustrates that some of the major uses of electricity are refrigerators and hot water heaters because they have high wattage and are used continuously every day. Other high-wattage appliances, such as toasters and hair driers, do not consume very large amounts of electrical energy because they are used only for short periods of time. Laptops use less energy than desktops. Power-saving options on computers save energy for either type of device.

Table 8-1 Electrical Energy Use in the Home

Appliance	Wattage Rating
Aquarium	50–1210
Clock radio	10
Coffee maker	900–1200
Clothes washer	350–500
Clothes dryer	1800–5000
Dishwasher (with/without dry)	1200–2400
Dehumidifier	785
Electric blanket (single/double)	60/100
Fans	
Ceiling	65–175
Window	55–250
Furnace	750
Whole house	240–750
Hair dryer	1200–1875
Heater (portable)	750–1500
Clothes iron	1000–1800
Light bulb (as rated)	60–150
Microwave oven	750–1100
Personal computer	
CPU (awake/asleep)	120/30 or less
Monitor (awake/asleep)	150/30 or less
Laptop	50
Radio (stereo)	70–400
Refrigerator (frost-free, 16 ft^3)	725
Televisions (color)	
19-in	65–110
27-in	113
36-in	133
53–61-in projection	170
Flat screen	120
Toaster	800–1400
Toaster oven	1225
VCR/DVD	17–21/20–25

(continued)

Table 8-1 Electrical Energy Use in the Home (Continued)

Appliance	Wattage Rating
Vacuum cleaner	1000–1440
Water heater (40 gallons)	4500–5500
Water pump (deep well)	250–1100
Water bed (heater, no cover)	120–380

Measuring Electrical Power and Energy

Electrical power is measured in watts (W). If you are dealing with a lot of watts, it may be easier to refer to kilowatts (kW): 1000 W make up 1 kW. (Additional information about units for power, energy, and other quantities used in this book can be found in the Appendix.). Power is not the same as energy. Energy is how much power you use for a given amount of time. Electrical energy is measured in kilowatthours (kWh). If you burn a 100-W light bulb for 10 hours, you consume more energy than if you keep it on for only 1 hour (Table 8-2).

Standby Power

Some studies have shown that the typical American home constantly wastes 20–60 W of electrical power. This is like never shutting off one small-wattage bulb. These phantom appliances include an average of 10 unneeded electrical devices on at any given time, including battery chargers, inkjet printers, DVD players, garage door

Table 8-2 Electrical Power and Energy

Formula for calculating electrical energy usage:
Watts × hours used ÷ 1000 = kilowatthours (kWh) electrical energy consumed
Example 1
A 100-W light bulb has 100 W (or 0.1 kW) of power.
If you leave it on for 1 hour, you consume 100 W × 1 h ÷ 1000 = 0.1 kWh.
If you leave that same light bulb on for 100 hours, you consume 10 kWh of electrical energy.
Example 2
A desktop computer and monitor are used 4 hours per day. How much energy is consumed in a year?
((120 + 150 W × 4 hours/day × 365 days/year) ÷ 1000 = 394 kWh)

openers, and ready-mode sound and video systems. Cable and satellite set-top boxes often consume more power than the television set they serve. A simple rule of thumb is that if it feels warm when it is presumably off, it is consuming power.

This wasted power adds up to $3.5 billion annually and accounts for the generation of almost 1 percent of total carbon dioxide emissions. Eliminating all this parasitic wasted electricity would save the average household 5–10 percent of its energy costs, resulting in a savings of about $400 annually per household (S. Hoffman, ElectricNet, www.electricnet.com). One simple way to eliminate much of this waste (besides unplugging unneeded devices) is to use a switchable power strip.

Although most of this waste is occurring in the United States, much of the electronic equipment involved is manufactured in developing countries. Since this is potentially a global problem, countries throughout the world can play a role in reducing this source of wasted electricity. Eliminating this phantom standby power represents the low-hanging fruit, but there are many other areas in which electricity can be conserved (as well as overall). Additional thoughts about conserving energy both at an individual and at a national level are listed later in this chapter.

Lighting Efficiency—How Many Scientists Does It Take to Screw in a More Efficient Light bulb?

A little over one-third of all energy is used to produce electricity. Slightly more than one-fifth of all electrical energy is used for lighting. Incandescent light bulbs are extremely inefficient, wasting over 95 percent of the electrical energy by producing heat rather than light. What makes matters worse is that heat generated by light bulbs constitutes a significant part of the cooling load for office buildings. Figure 8-3 compares the various types of light bulbs that are presently available or under development.

Fluorescent bulbs, including the screw-in type compact fluorescent bulbs, are five times more efficient than incandescent bulbs. Light-emitting diodes (LEDs) are, in a way, comparable with a solar cell in reverse. They convert electrical energy to light energy. LEDs have the potential to achieve 10 times better efficiency compared with incandescent and twice the efficiency of fluorescent bulbs. LEDs of various colors are now used commonly for electronic component indicator lights, outdoor displays, and automobile brake lights. White-light LEDs emit ultraviolet light on a phosphor that produces a mix of visible colors that people see white light. This technology has made inroads in initial markets such as high-efficiency flashlights and solar-powered walk lights. Several companies (including GE and Phillips) have been actively pursuing commercialization of white LED light bulbs (also referred to as *semiconductor lighting*). This represents a significant opportunity to conserve electrical energy and reduce greenhouse gas emissions.

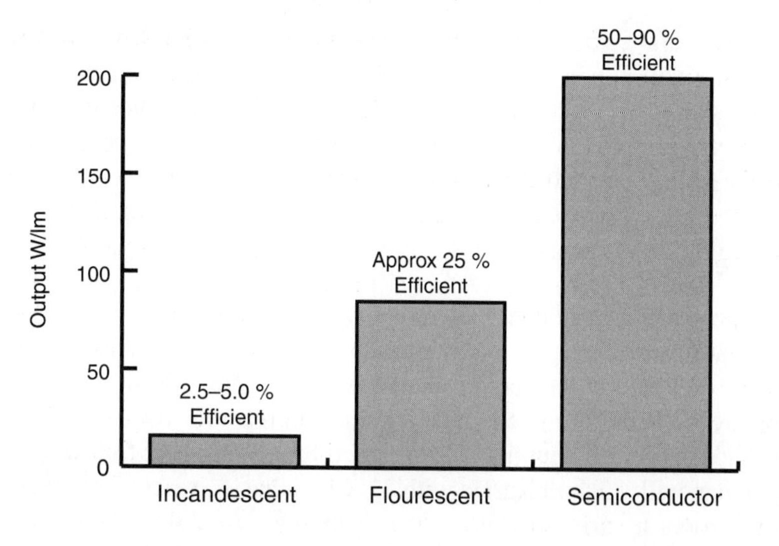

Figure 8-3 Light output for various types of light bulbs.

ALTERNATIVES TO COAL-GENERATED ELECTRICITY

Coal is not the only game in town. Coal currently may be one of the most convenient ways to produce electricity, but from a long-term environmental perspective, it is clearly not one of the best. At present, one-third of the world's electricity is generated by hydroelectric (water) power and by nuclear power. Neither is a source of greenhouse gases.

Hydroelectric

Hydroelectric (or *water*) *power* is a nonpolluting, non-greenhouse gas–generating source of electricity that currently is providing 16 percent of the world's electricity. Its application is very site-specific, and many of the good sites have been taken, with the balance of sites that remain to be exploited located in developing countries. The largest site in the world is the Three Gorges Project located on the Yangtze River in China. On its completion, it will generate nearly 20,000 MW of power. By comparison, the Hoover Dam in the United States and the Aswan Dam in Egypt have about 2000 MW of capacity. Two other major sites in South America have over 10,000 MW of capacity. According to some estimates, there is enough potential to increase the installed capacity of hydroelectric generating sites by a factor of 3 or 4.

Opportunities also exist for smaller hydroelectric sites. Concerns with impacts on wildlife and water resource management make it increasingly difficult to expand hydroelectric power to new sites. Hydroelectric power will be part of the mix, but it may not grow as rapidly as other renewable sources of power.

Exploitation of the energy from the oceans, both from wind-driven waves and ebb and flow of tides, represents a huge and mostly untapped potential renewable nonpolluting source of electricity. Ocean thermal electric conversion (OTEC) systems extract energy from the oceans by taking advantage of the temperature difference between the top surface and cooler water at depths below 1000 m (0.62 mile).

Nuclear

A nuclear power plant produces electricity by converting heat generated by splitting atoms into steam, which turns a turbine. Since nuclear power does not involve the combustion of a fuel, it does not produce greenhouse gas emissions.

There are 438 operating commercial nuclear reactors throughout the world. These are located in 30 countries. The United States has 104, followed by France with 59, Japan with 55, and Russia with 31. China has 11 reactors. A breakdown of nuclear reactors for various countries is shown in Figure 8-4.

Dozens of new reactors are currently being planned by China, India, and Russia. Several countries are currently exploring building their first nuclear reactor. Since 2000, more than 20,000 MW of nuclear capacity has come online throughout the world.

Nuclear power provides nearly one-sixth of the world's electricity, with larger percentages in some countries, such as France (78 percent), Belgium (54 percent), Sweden (48 percent), Hungary (38 percent), and Germany (32 percent). Nuclear

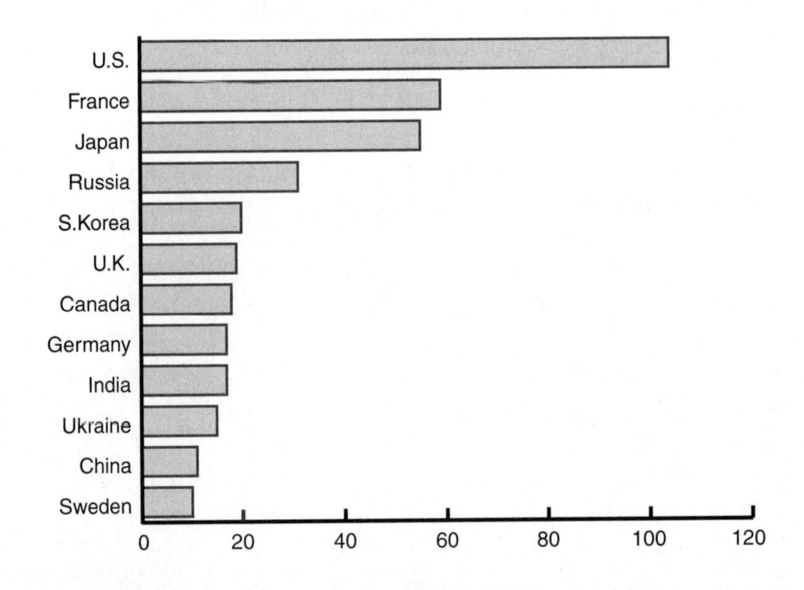

Figure 8-4 Number of nuclear reactors operated by various countries.

power provides 20 percent of U.S. electricity requirements. Despite hitting the early speed bumps of Three Mile Island in the United States and Chernobyl in the former Soviet Union, nuclear power has been reliable and efficient for the past several decades. According to the U.S. Energy Information Administration, nuclear reactors are now operating at nearly 90 percent of design capacity. This represents a significant increase from 1980 during which the operating capacity was only 56 percent.

No firm orders have been placed for new nuclear reactors in the United States for decades based mostly on concerns about waste management and maintaining costs. However, in the Far East, new nuclear power plants capable of providing 2000 MW of electrical power are coming online. Finland has taken the lead in setting up a site for its high-level nuclear wastes at an underground storage facility a half a kilometer underground. This is expected to be functioning by 2020. Most other countries are still seeking ways to store and, more important, secure spent nuclear materials.

The United States has not started construction on a new nuclear plant for three decades. The last new reactor to come online in the United States was the Watts Bar 1 in eastern Tennessee, which began operation in 1996. Construction had started on this site 23 years earlier. NRG Energy, Inc. recently submitted an application to build two new nuclear reactors in South Texas.

Nuclear power presently accounts for 16 percent of the world's electricity and 20 percent of the electricity in the United States. To promote nuclear power as a possible way to alleviate greenhouse gas emissions, in 2005, as part of the Energy Policy Act of 2005, the United States introduced new financial incentives (a credit of 1.8 cents per kilowatt hour) for new nuclear power plants during their first 8 years of operation. This may help nuclear power, which have an average electricity cost of 6.7 cents per kilowatt hour, to be more competitive with coal-powered plants, which have an average electricity cost of 4.2 cents per kilowatt hour. The nuclear industry conceivably would continue to benefit from economic incentives if carbon emissions are assigned a cost.

Nuclear power plants require cooling water to condense and reduce the temperature of steam that drives the turbines. This cooling water raises the temperature of the body of water into which it is discharged and has been described as thermal pollution. Typically, the maximum temperature of the discharged water is regulated. Site licensing delays have required resolution of this and other environmental impact issues. As climate change reduces available water resources in some areas, including those that will evolve toward greater drought conditions, nuclear power plants in those water-challenged areas will have a more difficult time operating. During the European heat wave of 2003, nuclear power plants in France were forced to shut down because of water restrictions. Nuclear plants in the United States and Japan have already had to limit operations because of water conditions that were not fully anticipated when the power plants were built. This came at a time of peak electrical demand from air-conditioning loads during the summer months.

There have been no major accidents since the two most severe accidents affecting this industry: Three Mile Island in the United States in 1979 and Chernobyl in the former Soviet Union in 1986. However, this growing technology has had numerous minor incidents that have not led to more serious problems. Since nuclear power plants have a 40- to 50-year design life span, the industry will have to address issues of component and structural breakdown and the economic costs of removing nuclear power from service.

Although nuclear power can be considered part of the solution to the global warming problem, it does place an enormous burden on those operating the nuclear power plants to keep the nuclear materials safe and out of the wrong hands. Needless to say, nuclear materials must be secured throughout the world regardless of whether the nuclear industry expands from its present level. Because it is a non-greenhouse gas–producing source of energy, we can expect nuclear power to continue to grow. The challenge will be to balance this growth with the need for nonproliferation of weapons that can be made from even small quantities of nuclear waste material.

Wind

Today, wind energy provides only about 0.5 percent of the world's electricity. But wind power has found a niche in parts of the world and has been growing at an average rate of 28 percent per year worldwide since 2000. In 2005, there was with a record increase of 40 percent in the use of wind-generated electricity owing to cost reductions and government incentives. Wind electric generators are two- or three-bladed propeller systems that are often clustered together in wind farms.

One possible silver lining of climate change is that higher average wind speeds are expected in some places, allowing wind-generated electricity to become more competitive. The trend over the last 25 years has been toward larger wind turbines. In the early 1980s, systems typically produced fewer than 50 kW. Today, the largest commercially available wind turbine has a rotor diameter of over 120 m (394 ft) and generates around 5 MW. The average today is in the range of 1.6–2 MW. (A *megawatt* is 1 million watts, which can provide the electrical power for roughly 1000 homes in the United States and a larger number of homes in other countries.)

Wind accounted for 18.5 percent of the electrical energy in Denmark, using installations such as shown in Figure 8-5, giving Denmark the highest usage per person anywhere in the world. (Western Denmark is setting the wind energy use record by producing 25 percent of its electrical power from the wind.) Wind energy provides roughly 15 percent of the electrical energy in Spain.

On good sites, well-designed systems can produce electrical power for around 3 to 5 cents per kilowatthour (in terms of U.S. currency), which is competitive head to head with fossil fuel–generated electricity. Wind-generated electricity can be used to produce electricity independently, to offset fossil fuel–generated power by

Figure 8-5 Wind energy farm providing electricity off the coast of Denmark. (*Source: US DOE; credit: Sandia Labs.*)

utilities, or connected to the power grid and sold to utilities. "For farmers, one wind turbine can rake in about $5,000 a year in rent, compared with $300 for corn or soybean farming."[2] Opposition to more widespread use of wind is in some locations can be impeded by aesthetic or other environmental concerns that result in a not-in-my-backyard syndrome.

Not all but many other selected locations around the world have an opportunity to cut their carbon dioxide emissions using wind energy (or "airtricity," as it is called by some people). A global study of 7500 sites showed an average wind speed at 80 m above the ground of 6.9 m/s. The greatest potential was found in northern Europe, the southern tip of South America, the Great Lakes Region, and the northeastern and western coasts of Canada and the United States. If fully exploited, wind power could provide five times the anticipated global electrical requirement in 2005. Much of the growth in wind power is expected to provide power to meet growing demand rather than to displace fossil fuel–generated electricity. Major investments in wind energy have occurred in Europe, Japan, China, the United States, and India.

Wind energy can be intermittent, with occasional periods when electrical output stops entirely. Connecting wind farms to a larger power grid enables utilities to use electricity generated by the wind when it is available without the need for dedicated storage.

---

[2]R. Wolverson. The anger Is Blowing In The Wind, *Newsweek*, November 12, 2007.

Solar Electricity (Photovoltaics)

Enough energy from the sun strikes the earth in 1 hour to provide all the energy consumed by the earth's entire population in 1 year. In most places, sufficient sunlight strikes the earth's surface to power an (energy-efficient) home and support a plug-in hybrid car. Solar energy stands out as an opportunity that overshadows all the other renewable energy sources and fossil fuels combined. Today, only a small fraction of this vast potential has been exploited.

Solar cells were developed (by accident) at Bell Labs and have been used to power most satellites launched since the 1950s. The development of *terrestrial* (earth-based) solar electricity began in response to the oil crisis of the early 1970s. Solar cells, or *photovoltaic* energy, convert energy from the sun into direct current electricity. There are no moving parts and no greenhouse gas emissions.

Photovoltaics are used to power nearly every satellite in orbit and have become a common source of energy for many calculators. It is a proven and reliable technology. Much of the commercial interest in photovoltaics over the past several years has been for remote power, such as to power roadside signs, call boxes, and buoys at sea.

Since an initial push in the 1970s, solar power found a niche to provide electricity where it was otherwise inconvenient, too expensive, or impractical to connect to existing power lines. Photovoltaics power refrigerators in remote villages to keep vaccines cold. They provide a small voltage to inhibit corrosion of pipelines and run irrigation pumps. Many of the photovoltaic installations over the past few decades were used to provide new power rather than to replace electricity generated by fossil fuels.

Today, the fastest-growing market for photovoltaics involves grid-connected applications that use an electronic component (called an *inverter*) to convert the direct current that the panels generate to alternating current. The power generated can be used to meet electrical needs of a home or business. If a solar array produces more power than is needed at a particular time, that power can be delivered to the power grid for distribution to other customers by the local utility. Individual producers of solar-generated electricity can (legally) run their electric meters backwards and be compensated for that power. Figure 8-6 shows a home in California that produces electricity using photovoltaic panels on its roof.

Although power purchased by utilities is regulated at the state level, it can provide a benefit to the utilities by providing power during peak-demand periods. During the hottest part of the day, when air-conditioning loads are greatest, local grid-connected photovoltaic systems can help utilities to avoid firing up their more costly supplemental natural gas–fired turbines. It also can reduce the need to add capacity just to meet peak demand.

During the past decade, photovoltaic installations have increased by 30 percent annually, with the trend going from isolated stand-alone applications to grid-connected

Figure 8-6 Roof-mounted photovoltaic system. In many locations, utility companies purchase solar-generated electrical power from customers. (*Courtesy: BP Solar.*)

systems. In 1994, 60 percent of the photovoltaic market was remote power. Eighty percent of the systems installed in 2004 were of the grid-connected type. Much of the growth of this technology is the result of financial incentives provided by Germany, Japan, and states in the United States, including California, Arizona, and New Jersey. Today, nearly 50 percent of the new photovoltaic installations have been in California. The very rapid pace of the growth of photovoltaic systems in these areas can be seen in Figure 8-7. The total installed capacity is approaching 4000 MW worldwide.

The major obstacle to widespread use of solar electricity is its greater cost relative to fossil fuels. Presently, photovoltaics are about three times more expensive than they need to be to compete in today's energy market with fossil fuel–generated electricity.

There are three main types of solar-cell technology:

1. *Silicon wafers.* Most commercial solar cells use silicon as the base material that absorbs light energy and converts it to electricity. The silicon can be either crystalline (as is used in Silicon Valley electronic applications) or a slightly less perfect but less expensive *semicrystalline* form. Silicon wafers represented more than 90 percent of the solar-cell market in 2004. Solar-cell manufacturers have been making use of excess or slightly off-grade

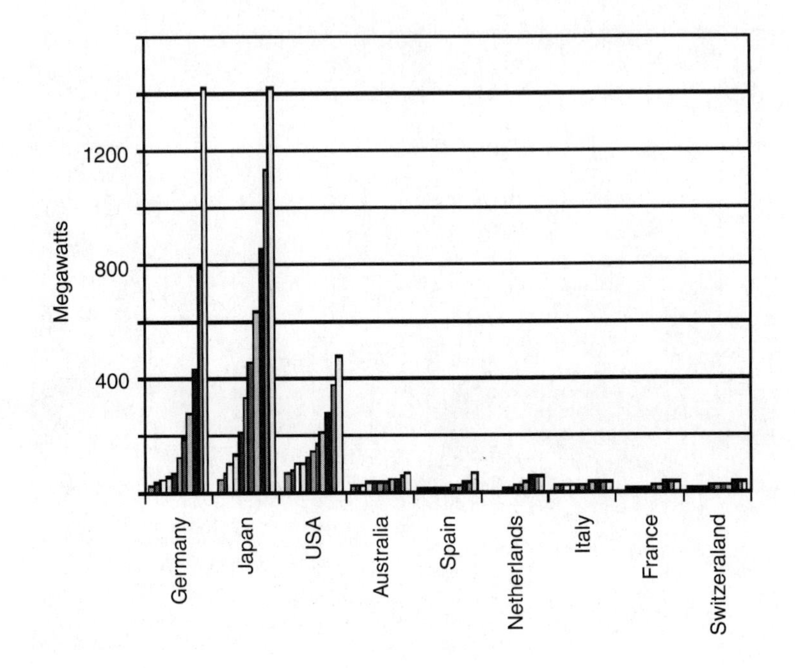

Figure 8-7 Cumulative installed solar electric power by various countries from 1995 to 2005.

silicon wafers intended for the semiconductor electronics industry. As a result of the surge in photovoltaic sales, the industry must cope with a shortage in silicon. Other approaches are to produce silicon in flat sheets to avoid the expense of cutting.

2. *Thin film.* Certain materials such as cadmium telluride, copper indium gallium diselenide, and noncrystalline (amorphous) silicon absorb sunlight using very little material. This provides a potential cost savings, but often at the expense of performance. Often there is a large gap between best laboratory proof-of-concept results and actual performance achievable in the field. Thin-film photovoltaics represents just under 10 percent of the solar electric market today.
3. *Concentrating photovoltaics.* This technology focuses the incoming sunlight onto small areas of high-performance solar cells. Since the focusing materials can be much less expensive than the active solar-cell area, use of concentrating solar system represents another approach to cost reduction. The difficulty is that concentrating systems often need a way to at least partially track the sun, which adds to the complexity and cost of the system. Concentrating solar systems remain an area of active interest but have not achieved commercial status yet.

Achieving cost targets for solar electric systems requires reducing the cost not only of the solar cells but also of other aspects of the system. Although the system contributes much to the overall cost of photovoltaics, the cost of the solar cell is a closely watched metric as to its commercial readiness. Currently, electricity generated by solar cells (averaged over the useful lifetime of a system) costs in the range of \$6 per watt. Bringing the cost down to \$4 per watt is seen as the goal to make photovoltaics competitive on their own. In some situations, development is needed in the other "stuff," or the balance-of-system costs. This applies to concentrator systems that have greater cost in components that focus or concentrate light to enable a much more expensive but state-of-the art solar cell. When a cost study indicates that the photovoltaic system is not cost-effective, even with zero-cost solar cells, it becomes clear that the next effort needs to go into overall system costs.

Opportunities exist to reduce cost in all aspects of the system over its useful life. The industry has been following a learning curve that has resulted in a decline of 20 percent in the price of photovoltaics for every doubling of worldwide production. In the mid-1950s, when the first commercial solar cell came out, it sold for nearly \$1800 (in 1955 dollars). It was 1000 times more expensive than fossil fuels. In the 1970s, costs had come down, but photovoltaics still were 100 times more expensive than conventional electricity. Today, the difference is in single digits, with the playing field being leveled by the support of governments around the world. Assigning a cost to carbon dioxide emissions may close the gap in commercial markets to allow solar electricity to continue to play a greater role in the future.

At present, worldwide photovoltaic generating capacity is expected to exceed 1 GW, or enough to power several hundred thousand homes (Figure 8-8).

Figure 8-8 Vallejo photovoltaic installation. (*Courtesy: BP Solar.*)

Solar Heat

As long as the sun is going to the trouble of heating up the earth, we may as well take advantage of its efforts wherever we can. Solar heat, or solar thermal energy, reduces the demand for fossil fuels or electricity generated by fossil fuels. There are three main methods for doing this:

1. *Passive solar energy for buildings.* By planning to use available incoming solar energy, a building can reduce its heating requirements significantly without much added cost. Passive solar design includes use of south-facing windows and overhangs that are sized to allow sun to enter the building in the winter but that will block the summer sun. Use of high-heat-capacity materials in the area where the sun strikes provides some storage of heat during the most intense periods of sun and minimizes the likelihood of overheating. Passive solar design usually makes the most sense in new construction and is most effective in well-insulated structures.
2. *Active solar energy for buildings.* Commercially available solar heating systems that circulate a fluid (usually water or air) through solar collectors mounted on roofs or in yards can contribute to the heating needs of a building. The fluid is driven by a pump or fan, with heat either going directly into the building or warming a storage medium such as rocks or water. For many parts of the world, this will be a supplement to a primary system that does not use solar energy. China currently leads the world in making use of solar heating, with an estimated 80 percent of all installations worldwide. The Chinese solar heating panels are made from evacuated tubes rather than the flat-plate designs used elsewhere.
3. *Hot-water heating.* Domestic hot-water heaters are probably the simplest and most widespread application of solar energy. These are well suited to warmer locations, where added costs are not needed to provide for freeze protection.

Geothermal Energy

Geothermal energy refers to producing heat from the earth. Theoretically, this potential resource alone can supply all the world's energy needs if fully exploited. Some geothermal sites are near the surface and are readily accessible. Other sites require drilling to the layers of heated rock 10 km (6.2 miles) or more beneath the earth's surface. At least 20 countries around the world are using geothermal energy, including Iceland, the United States, Italy, France, New Zealand, Mexico, Nicaragua, Costa Rica, Russia, the Philippines, Indonesia, China, and Japan. Kenya will soon be able to provide close to one-quarter of its electrical requirements using geothermal energy.

Geothermal electricity is generated by steam produced from underground heat turning a turbine. If the underground heat is not hot enough to produce steam on being brought to the surface [182°C (360°F)], the water in the geothermal reservoir is passed through a heat exchanger that transfers the heat to a separate pipe containing fluids with a much lower boiling point. Systems using this form of heat exchange (known as *binary-cycle plants*) have the advantages of lower cost and increased efficiency. Most geothermal power plants planned for construction are binary-cycle plants.

An interesting tradeoff to explore is whether it is more cost-effective to retrofit an existing coal-fired electricity plant to run off geothermal power rather than to go to the effort of capturing and storing the carbon dioxide emissions.

The temperature of the earth slightly below its surface is close to 55°F (13°C). Geothermal energy also can be used for heating homes directly, as is being done in about 30,000 locations in Canada. Use of an underground heat sink (instead of colder winter air) enables electric heat pumps to be much more efficient. This approach also can be used for cooling in parts of the world where it is needed.

Transportation—The Problem with Oil

AUTOMOBILE EFFICIENCY—HOW MUCH OIL IS REALLY NEEDED TO MOVE JUST ONE PERSON?

One-quarter of the greenhouse gases generated throughout the world are a by-product of transportation. In the United States, each person consumes, on average, 1.3 gallons of gasoline each day. This is 10 times greater than the average for the world. America continues to exhibit an infatuation with the automobile. Just as adolescents in industrialized countries count the days until they are able to drive, a similar interest in becoming mobile exists, and understandably so, throughout the emerging world.

Today, more than 7 out of 10 people in the United States own cars. In Europe, this number ranges from between 2 and 5 people out of 10. In the rapidly developing countries of China and India, that number is much less than 1 person out of every 10. If the rest of the world consumed gasoline at the rate that it is consumed in the United States, 10 times the current amount of greenhouse gases would be generated.

Standard spark-ignition internal combustion engines today are perhaps 35 percent efficient under ideal conditions and 10–20 percent efficient in typical urban driving. Considering that in today's cars roughly 300 pounds of people and gear are moved through traffic in a vehicle that is 10 times that weight, the overall efficiency of the fuel in performing its primary function is only about 1–2 percent. From this, we can take heart in how much room there is for improvement.

GENERATING LESS CARBON DIOXIDE

Gas–Electric Hybrids

Hybrid vehicles have two motors—an electric motor powered by a battery and a significantly downsized gasoline engine. The battery is charged by the operation of the gasoline engine and by *regenerative braking*, which recaptures mechanical energy that otherwise would be lost during braking.

Hybrid designs save energy in the following ways:

- They use the electric motor to provide power during acceleration. This enables the gasoline engine to be much less massive.
- When the vehicle brakes, the energy is recovered as electrical energy, which is then stored in the battery.
- Shutting down the engine when the vehicle is stopped eliminates waste. This is especially true during city driving, where there can be a lot of unproductive idling time in traffic.
- Use the electric motor instead of the gasoline engine at slow speeds eliminates engine operation when it is least efficient (Figure 8-9).
- Power steering and other accessories can be shifted to more efficient electrical operation.

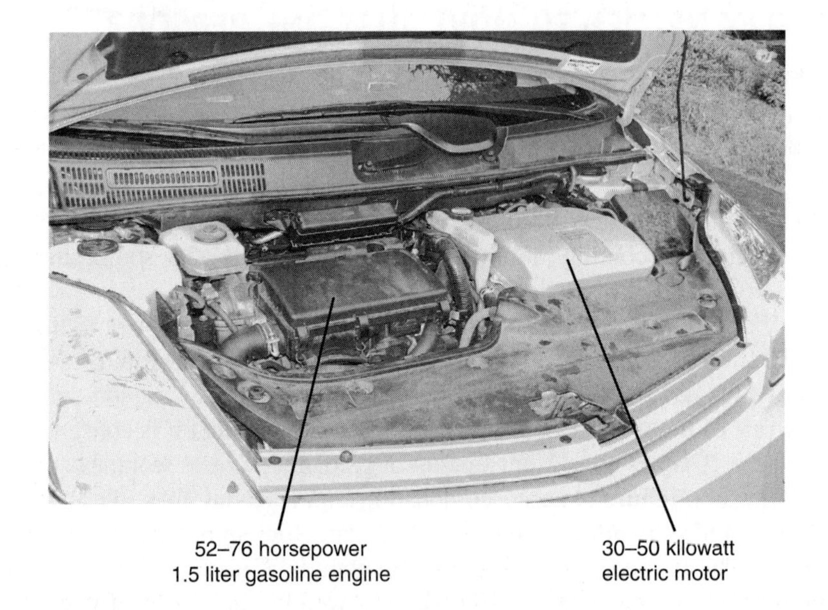

Figure 8-9 Under the hood of a hybrid car. An electric motor allows the use of a much smaller and more efficient gasoline engine.

Transmission improvements also reduce loss of power. Use of nickel–metal hydride batteries similar to those used in satellites instead of lead–acid batteries increase the amount of energy a battery can hold and enhances its performance. A hybrid, such as the Toyota Prius, just about doubles the fuel economy of comparable nonhybrid cars, with even better numbers under congested urban conditions.

How much will this help? Let's assume that 2 billion cars are expected to be on the world's roads by the middle of the twenty-first century and that all achieve 60 miles per gallon (mpg) instead of the 30 mpg or so that is typical of today's cars. This will prevent 1 billion tons of greenhouse gas emissions each year (see the section "Steps Toward a Solution" later). Of course, there are differences in how automobile gasoline mileage is measured. More "realistic" driving conditions result in lower numbers. Still, the best-performing gas–electric cars such as the Prius show a factor of 2 improvement over comparable cars that do not have an electric motor. Moreover, improvements in how much energy can be stored in a given battery mass (called the *specific power of the battery*) can be expected to translate directly to improved hybrid car performance.

Plug-in Hybrids—Producing Much Less Carbon Dioxide

How about taking a hybrid car and putting in an even higher-performance battery that can be charged when the vehicle is not in operation? A supersized battery would reduce the need for the gasoline engine substantially. Basically, this would be an electric car with a small gasoline motor to supplement the power of the electric motor the small percentage of the time that greater acceleration is needed. The gasoline motor also would extend the range of the vehicle so that it could continue back home or to the nearest place to recharge the battery if the battery runs out. Mileage approaching 100 mpg is reasonable to expect from this approach and since most car owners drive less than 35 miles (16 km) per day recharging overnight may be very feasible. Mileage in the range of 100 mpg can be accomplished on some current commercially obtained hybrid cars by using a pre-charged lithium ion battery rather than the standard nickel-metal hydride battery *(http://auto.howstuffworks.com/100-mpg-news.htm)*.

In a Prius "hypermileage" marathon, a team drove 1397 miles along a 15-mile course on 12.8 gallons in just under 48 hours. The result: an average 109.3 miles per gallon setting an unofficial world record. 120.6 mpg was established on the best segment of the course. The improved mileage came from driving techniques that minimized the amount of time that the engine runs and minimizes power flowing to and from the battery. Although these techniques are not practical for most driving conditions, it serves as a proof of concept that significantly higher mileage is possible *(www.toyota.com/html/hybridsynergyview/2005/fall/marathon.html)*.

While operating by the electric motor, plug-in hybrids would not generate any greenhouse gases or any form of pollution, for that matter. However, since plug-in hybrids get some of their energy from the electrical power grid, the opportunity to reduce greenhouse emissions depends on the sources of the electricity used to charge the battery. If the electricity comes from a coal-fired electrical power plant that does not capture and store its carbon dioxide emissions, the benefit of the cleaner operation of the plug-in hybrid would be greatly diminished. If the electricity is produced without carbon dioxide emissions, the impact of the plug-in vehicle can be significant in addressing the transportation part of this problem.

> Automobile designs that use precharged batteries (such as plug in hybrids) contribute to reducing global warming only to the extent that the electricity they use to charge their batteries does not generate greenhouse gases.

CLEANER FUELS

Ethanol—Corn, Sugar, Cellulose

Ethanol, or grain alcohol, can be produced from corn, sugar beets, sugar cane, or other crops primarily by fermentation. Ethanol came onto the scene largely as a means of moving toward energy independence. Brazil currently uses ethanol to meet an estimated 40 percent of its transportation requirements. Presently, roughly 20 percent of the corn grown in the United States is converted to ethanol. Current farming methods use a high percentage of petrochemicals, which to some extent defeats the intent of displacing oil.

Mixtures of ethanol and gasoline (such as 15 percent ethanol and 85 percent gasoline, or E85) are becoming common as an alternative fuel in certain areas of the United States. The energy payback from corn-grown ethanol, however, is marginal. Depending on agricultural conditions, ethanol produces on average only 25–30 percent more energy than the energy it took to produce it. This results in a net energy benefit, with the actual numbers depending on the specific production conditions, including how much carbon dioxide emissions are needed grow, harvest, process, and distribute the ethanol. A higher percentage of ethanol in the fuel blend may require engine modifications. Some automobile manufactures are now offering *flexible fuel vehicles* (FFVs) to accommodate either gasoline or higher-percentage ethanol mixes. Because of its chemical structure (carbon–oxygen bonds rather than the more energetic carbon–hydrogen bonds found in petroleum-based fuels), ethanol delivers about 30 percent less energy per gallon than gasoline. This may not be as noticeable with low-ethanol blends but may become more of an issue when there is more ethanol in the mix.

Since ethanol contains carbon in its chemical structure, it, like any other carbon-containing fuel, produces carbon dioxide when burned. For a given amount of energy produced from the same size fuel tank, both ethanol and gasoline produce comparable amounts of carbon dioxide. One difference is that the carbon dioxide that gets released to the atmosphere when ethanol is burned came from the atmosphere through the process of photosynthesis that produced the corn. Ethanol can be thought of as just returning the carbon it removed from the atmosphere. This does not give ethanol a real advantage over gasoline, however, because if the corn wasn't removing carbon dioxide, presumably some other crop would be there in its place. Growing corn to produce fuel requires farmland that otherwise could grow food crops. This could introduce price pressure on food at a time when increasing flood and drought conditions might diminish the usefulness of some agricultural areas around the world.

The process of producing ethanol involves a fermentation step that produces carbon dioxide. For every 0.51 kg of ethanol produced, 1 kg of carbon dioxide is produced. Capturing this carbon dioxide would improve ethanol's effectiveness in terms of greenhouse gas reduction. Additional carbon dioxide is released when ethanol is burned, but it is made up of the same carbon atoms that were removed from the atmosphere to grow the feedstock to produce the ethanol.

Brazil has pioneered the use of ethanol. The government mandated 25 percent use of ethanol as a means toward energy independence. Government support, available agricultural acreage, and a climate conducive to growing sugar beets helped to promote this effort. Whether or not other countries can replicate Brazil's experience with ethanol, the experience today does serve as a success story in implementing a change in a country's approach to energy production and use.

While achieving the twin goals of energy independence and pollution reduction, it is questionable whether ethanol can make much of a dent in the level of carbon dioxide emissions in the short term. Potential efficiency improvements in the ethanol growth and production cycle may improve this situation, this especially if organic farm wastes such as corn stalks, grasses, wheat and rice straw, leaves, and other agricultural leftovers (called *lignocellulosic* materials) are used as a starting material.

Cellulosic crops are attractive because they have higher yields than high-carbohydrate crops such as corn and sugar beets. They grow more easily in areas that are not suitable for grains or other food crops without the need for extensive fertilization. They do not necessarily compete with crops grown for human or animal consumption. The cellulosic materials can provide some of the process heat needed to separate the ethanol after the fermentation process, avoiding the need for consuming additional fossil fuels in the process. At this point, substantial more research is needed to make this a commercial option.

Biodiesel

When Rudolph Diesel introduced the engine that now bears his name at the 1900 World's Fair in Paris, he used peanut oil as its fuel, which by today's standards could be considered *biodiesel* rather than *petrodiesel.* Thus the current resurgence of interest in the use of organic sources of fuel for diesel engines brings us full circle. Diesel engines are 30–40 percent more efficient than the more common internal combustion engines that use a spark plug to ignite a gasoline–air mixture. A diesel engine uses the heat generated by compression of the fuel mixture to produce the combustion that drives the engine.

Diesel engines are far more popular in Europe today than in the United States, where they are used commonly in buses and trucks. Throughout their life cycle, biodiesel fuels produce 60–75 percent less carbon dioxide than an equivalent amount of gasoline.

One of the things that makes biodiesel attractive in a way that does not apply to ethanol is that biodiesel can be made from recycled oil that has been used for cooking. The amount of waste oil available for this purpose, however, would be insignificant compared with the amount needed. Biodiesel appears less promising, however, than cellulosic ethanol in terms of cost and potential for commercialization.

In general, biofuels are limited by the amount of farmland around the world that can be dedicated to energy crop growth and by the availability of agricultural waste streams that can be converted to biofuels.

Hydrogen

From the point of view of global warming, hydrogen is perfect as a fuel. Hydrogen fuel produces no carbon dioxide and releases only water to the environment. Whether the hydrogen is burned in an internal combustion engine or reacts chemically in a fuel cell to produce electricity, it is by far the cleanest fuel imaginable.

Three things need to happen to make hydrogen fuel a reality:

1. *An efficient and cost-effective method to generate hydrogen must be developed.* The source is no problem. Hydrogen can be produced by passing an electric current through water. However, if the energy used to separate the hydrogen comes from a coal-fired electrical power plant, it defeats the purpose of using hydrogen in the first place. Use of a renewable form of energy such as wind or solar electricity would result in a net reduction of carbon dioxide emission. (Another way to produce hydrogen is to separate the hydrogen atoms in methane.)
2. *Hydrogen would need to be transported to where it is to be used.* If, for instance, gasoline trucks are used to bring the hydrogen from a separation plant to a "hydrogen filling station," as much carbon dioxide might be released

to the atmosphere as might have been saved by using hydrogen in the first place. Local generation of hydrogen close to its point of use is a better option but one that would requiring modifying the way gas stations work today.

3. *Finally, an infrastructure of filling stations would need to be established throughout the transportation system.* Technical problems would need to be addressed, such as the fact that hydrogen leaks much more easily than other gases such as natural gas and would need more robust containment and distribution systems.

Hydrogen-powered vehicles, including cars and buses, have been developed and have been proven to be technically feasible. They are ideal for the environment. However, the best estimates for the infrastructure to support a hydrogen economy are decades away. A lot of carbon dioxide will be generated in the next several decades in the meantime.

LAND USE

IPCC estimates put carbon dioxide emissions from deforestation, including decomposition following logging operations, to be between 7 and 16 percent of the world's contribution in 2004. Natural contributors to greenhouse gas production and sinking are larger than the added contributions from fossil fuel combustion and other human activities. Every year, a large amount of carbon dioxide (roughly 100 billion metric tons) is removed from the atmosphere and stored in plants and soil. Removal and release of carbon dioxide are roughly in balance worldwide. The U.S. Department of Energy estimates that plants absorbed 17 percent of the carbon dioxide produced by burning fossil fuels in 1992.

Forests hold an enormous reservoir of carbon. The U.S. Forest Service estimates that forests in the United States hold 56 billion metric tons of carbon, equivalent to nearly 40 years of emissions from fossil fuel combustion. Overall, U.S. forests, just as those in other countries, have been a net carbon sink in recent years.

METHANE

Natural gas. Methane emissions come from several natural and human-contributed sources. Reducing the human component centers on several industries. Since methane is the main constituent of natural gas, greater care during the production, processing, transmission, and distribution of natural gas will result in a lower level of emissions.

Petroleum. Crude oil production releases methane through venting from storage tanks and other equipment. This presents an opportunity to reduce emissions. Since methane is a fuel, if it can be captured in useful quantities, it could offset in part the cost of collecting it. Each molecule of methane that burns (completely) releases one molecule of carbon dioxide, which, as we know, is also a greenhouse gas. However, the ability of the carbon dioxide molecule to absorb energy is far less than the methane molecule. For this reason, given the choice of burning methane or releasing it, it is preferable to burn it.

Coal. Venting and possible reuse of methane captured from underground or surface coal mines is a way to reduce methane emission.

Agriculture. Improved feeding practices, such as using concentrates to replace foraged food and adding oils to the diet of livestock, can cut down on the methane produced on farms.

Landfills. Reducing the amount of waste that is brought to landfills is a good step toward reducing the release of methane. Collecting the methane generated and using it as a fuel, if possible, or burning it, if necessary, would cut down on methane release from landfills.

Steps Toward a Solution

It is inevitable that greenhouse gas emissions will continue to rise over the next several decades as the world grapples with an appropriate response. The question is the point at which the emissions are stabilized. Robert H. Socolow and Stephen W. Pacala[3] suggested that doubling of the carbon dioxide above preindustrial levels would be a reasonable "boundary separating the truly dangerous consequences from the merely unwise." They define two scenarios:

1. Emission levels continue to grow at current rates for the next 50 years, reaching 14 billion tons of carbon by the year 2056.
2. Emission levels are frozen at 7 billion tons a year for the next 50 years (and then reduced by half over the next 50 years).

One way to define an effective solution is to identify actions that will bring the world from the first scenario above to the much more benign condition represented in the second scenario. This gives us a better idea of what it will actually take to have a meaningful impact on the problem of global warming. To stabilize carbon dioxide emissions at current levels, it would be necessary to emit 7 billion tons a

---

[3]Robert H. Socolow and Stephen W. Pacala *Scientific American, September 2006,* A Plan to Keep Carbon in Check, p50.

year *less* than current levels for the next 50 years. Such an action likely would stabilize greenhouse gas concentrations well below 560 ppm (anticipating substantial absorption of the increased emissions by the oceans). Each of the following actions independently would prevent the release of 25 million tons of carbon if phased in over the next 50 years. One or two of them alone is not enough; it will take seven of these steps (or their equivalent) worldwide to stabilize greenhouse gas levels. Any seven of the actions (or combination) from the following list would result in that stabilization.

1. Increase average automobile mileage from 30–60 mpg—for 2 billion drivers.
2. Reduce average automobile driving distance from 10,000–5000 miles per year—for 2 billion cars.
3. Reduce worldwide electricity use by 25 percent.
4. Improve the efficiency of at least 1600 large coal-fired electricity-generating plants by from 40–60 percent.
5. Replace 1400 large coal-fired plants with natural gas–fired plants.
6. Install carbon capture and storage systems at 800 large coal-fired electricity-generating plants.
7. Install carbon capture and storage systems at coal plants that produce hydrogen for 1.5 billion vehicles.
8. Install carbon capture and storage systems at coal-to-syngas plants producing 30 million barrels of syngas daily.
9. Double the amount of nuclear-generated electricity to replace coal.
10. Increase the use of wind-generated electricity by a factor of 40 to replace coal.
11. Increase photovoltaic power generation by a factor of 700 to replace coal.
12. Generate enough hydrogen by increasing wind-generated electricity by a factor of 80 to produce hydrogen for cars.
13. Drive 2 billion cars on ethanol (Note: using one-sixth of the world's farmland and assuming substantial reductions in the carbon footprint of producing and transporting ethanol compared with ethanol produced today from corn).
14. Stop all deforestation.
15. Expand conservation tillage to 100 percent of cropland (growing crops without first tilling the soil).

These steps include a broad range of options that will stabilize and potentially reverse the climate changes that have been set in motion.

To stabilize greenhouse gas concentrations below 500–600 ppm, substantial reductions in carbon dioxide emissions from coal-generated electricity generation and internal combustion engines will be needed. This will not be achieved by a series of well intended gestures on the part of individuals. Instead, the world must fundamentally rethink how it produces and uses energy.

Stepping Up to the Plate—Taking Action

THE KYOTO PROTOCOL

The *Kyoto Protocol* was negotiated in Kyoto, Japan, in December 1997. The 169 countries that ratified this protocol have committed to reducing their emissions of carbon dioxide and five other greenhouse gases. The United States and Australia were the only industrialized countries at the time that did not ratify this treaty although Austrialia eventually signed the accord at a follow-up conference in Bali in 2007. China and India did ratify the Kyoto protocol but are not required to reduce carbon dioxide emissions under the present agreement.

The objective of the protocol is the stabilization of greenhouse gas concentrations in the atmosphere at a level that would prevent disruption of the climate system.

ACTIONS TAKEN BY VARIOUS COUNTRIES AROUND THE WORLD[4]

European Union

The European Union has committed to reduce greenhouse emissions by 8 percent below 1990 levels by 2008–2012 as their Kyoto target. An emissions trading scheme was put in place that applies mandatory carbon dioxide limits for 12,000 sites throughout Europe. Incentives are being provided to increase the use of renewable sources of energy. Agreements have been established with automakers to reduce carbon dioxide emissions of new cars by 25 percent below 1995 levels. Encouraged by government incentives, people in Germany installed 100,000 solar systems in 2006, representing 750 MW of solar electric generation. Approximately 50 billion kWh of power, providing 10 percent of Germany's needs, now comes from renewable sources. A total of 78 percent of France's electricity now comes from nuclear power that does not produce greenhouse gases.

---

[4]Data derived from *Climate Change 101: Understanding and Responding to Global Climate Change,* published by the Pew Center on Global Climate Change and the Pew Center on the States, www.pewclimate.org.

United Kingdom

The United Kingdom established a national target that is 20 percent below 1990 levels, and this exceeds the requirements of the Kyoto agreement. The Government has placed a tax on fossil fuel–based electricity for large users, and most of the revenues to be collected will be used for energy research. A target of 10 percent of electricity to be generated from renewable sources by 2010 was established.

Japan

Japan's Kyoto agreement is to reduce emissions by 6 percent. Separate agreements target reductions to 1990 levels for a major industry association and 20 percent below 1990 levels for a power-generating group.

China

China has established fuel economy standards that require all new cars and light trucks to achieve 19–38 mpg (depending on the class of vehicle) by 2005 and 21–43 mpg by 2008. China is working to improve the amount of energy used in relation to its gross national product (measured as its *energy intensity*) (Figure 8-10).

Figure 8-10 China intends to raise its energy intensity by 20 percent from 2006–2010 and by a total of 50 percent from 2000–2020. Pictured here is a highway to Shanghai. (*Photo courtesy of Donald Liebner.*)

China's national targets for renewable energy are for 15 percent of overall energy and 20 percent of electricity by 2020. Specific goals have been established for wind power, biomass, and hydroelectric power.

India

Efforts are underway to improve the efficiency of the electrical power sector. There is an effort to move toward larger, more efficient power plants. A goal of 10 percent of new power generation by 2010 has been established as India moves ahead with electrifying 18,000 rural villages. Biomass, solar, wind, and hydroelectric power are being considered to meet the growing demand. India is also in the process of converting taxis, buses, and other vehicles from gasoline to natural gas.

The United States

As a contributor of 25 percent of the world's greenhouse gases, the United States has a large opportunity to help stabilize the world's climate. The United States has contributed a great deal to the world's understanding of climate through research and monitoring efforts. In the United States, state and local governments have taken the lead in establishing emissions reduction programs in the form of regional alliances. The states have a great deal of authority to regulate electrical power and can play a significant role in bringing about change.

In a landmark decision in 2006, the U.S. Supreme Court decided that excessive carbon dioxide added to the atmosphere is a pollutant and can be subject to pollution-control laws. This provides a legal foundation for governmental regulation of greenhouse gas emissions in the United States.

Northeast Regional Greenhouse Gas Initiative

The governors of seven Northeastern and Mid-Atlantic states established a cap and trade program intended to cut carbon dioxide emissions from power plants in the region. Under this arrangement, credits can be used outside the electricity industry to provide greater flexibility in meeting targets. A regional database will be set up to monitor progress.

Western Governors' Association

Eleven western states have committed to strategies to increase energy efficiency, expand the use of renewable energy sources, and provide incentives for carbon capture and storage.

Additional State Efforts

The Southwest Climate Change Initiative (organized by Arizona and New Mexico), the West Coast Governor's Global Warming Initiative (organized by Washington, Oregon, and California), the New England governors and eastern Canadian premiers, and Powering the Plains (organized by the Dakotas, Minnesota, Iowa, Wisconsin, and the Canadian Province of Manitoba) all have begun similar efforts to reduce greenhouse gas emissions.

Iceland

Like France, with almost 80 percent of its electricity requirements supplied by a non-greenhouse gas–producing source (i.e., nuclear), Iceland is rapidly approaching the goal of energy independence. Iceland is unique in that it sits on rock of volcanic origin and has access to virtually unlimited geothermal energy. Iceland has exploited this natural resource, as well as abundant hydroelectric sites, to provide about 70 percent of its energy needs—from home heating, to electricity generation, to industrial applications.

Neither geothermal energy nor hydroelectric power can power Iceland's cars and trucks. Iceland has an official national goal of converting all cars, buses, trucks, and ships to hydrogen by 2050. The world's first hydrogen filling station, run by Shell, opened in Reykjavik in April 2003. To offset the cost of hydrogen relative to gasoline, Iceland is hoping to be able to at least partially use geothermal energy to produce hydrogen fuel.

THE UNITED NATIONS CLIMATE CHANGE CONFERENCE IN BALI, INDONESIA

In December 2007, the member nations of the United Nations met to formulate a plan to reduce greenhouse gas emissions and to continue commitments made by participants at the Kyoto conference in 1997. The Kyoto Protocol had committed 36 industrialized nations to reduce greenhouse emissions by an average of 5 percent between 2008 and 2012.

Based on modeling results, the IPCC proposed a worldwide stabilization level of 445 part per million carbon dioxide to prevent the earth's temperature from rising more than 2°C (3.6°F) above pre-industrial levels. This would enable the world to avoid the most severe impacts of global warming such as drought, failed crops, increased hunger, inundation of small island countries, and widespread extinction of species. To achieve this stabilization level, the IPCC scientists and climate modelers proposed a worldwide reduction in greenhouse gases in the range of 25–40 percent by 2020.

Issues

The Bali conference provided an opportunity for countries around the world to address obstacles that previously stood in the way of greater cooperation in fighting global warming. Some of the developing countries, particularly China, wanted to see the United States, which at the start of the conference was the only industrialized country not to have signed the Kyoto treaty, to take a greater responsibility for reducing global carbon dioxide levels. Much of the carbon dioxide that had been released into the atmosphere so far has contributed primarily to the prosperity and high standard of living in the United States which stands out as having one of the higher per capita generations of carbon dioxide in the world.

The United States was concerned that actions it was being asked to take to reduce carbon dioxide emissions would slow its economic growth. The United States was also concerned that if it alone cut greenhouse gases, that effort would not be effective without similar actions in the rapidly developing countries. The developing countries—especially those with the largest populations, most notably China and India—are currently experiencing rapid economic growth which is highly dependant upon continued combustion of coal. Countries striving to achieve a higher standard of living for their people are reluctant to take on the added burden of cutting greenhouse emissions without their counterparts in industrialized countries accepting a comparable role.

Results

Consensus

After marathon negotiations that appeared on the brink of collapse several times an overall consensus was reached. The plan establishes in principle that "deep cuts in global emissions will be required" and provides a timetable for two years of talks to provide the first formal addendum to the 1992 Framework Convention on Climate Change treaty since the Kyoto Protocol 10 years ago.

At Bali, the world's nations including the United States agreed to negotiate on a deal to tackle climate change. Developing nations—particularly growing economies like China and India—committed to "measurable, reportable, and verifiable, nationally appropriate mitigation actions."

The Bali agreement initiated a two-year United Nations-sponsored process, intended to produce a binding international climate pact by the end of 2009. This could change the way industrialized and emerging nations work together to preserve a rapidly warming Earth. However, the agreement in Bali postponed many tough decisions and stopped short of the more aggressive and specific emission reduction targets advocated by the European Union and others. There is also no language making specific emission reductions mandatory. The conference ended in the adoption of the *Bali roadmap*, which sets a course for a new negotiating process to be concluded by 2009 leading to a post-2012 international agreement on climate change.

Deforestation/Reforestation

The Bali conference included provisions for international projects to limit deforestation and to restore forests where they had previously been destroyed. This can enable deforestation projects to attract money from private investors interested in storing up credits that can be redeemed at a higher price in future. Credits from avoided deforestation will be stored up in the same way as credits from renewable energy projects as part of the global market in carbon. Part of the financing would come from developed countries through aid. Additional financing would come from carbon credits traded under the Kyoto pact. Rain forest destruction is a major source of carbon dioxide and living rain forests play an important role in absorbing the gas. For this to be meaningful, it will be necessary to insure that projects will help reduce overall emissions instead of just push more deforestation elsewhere.

Adaptation Fund

One specific accomplishment in Bali was to implement the climate change adaptation fund. This fund, which was an important feature of the Kyoto Protocol, intended to help developing nations to adapt to the more-frequent, more-intense droughts, increasingly severe storms, and sea-level rise, that scientists project will occur as the planet's atmosphere warms. The climate change adaptation fund will also be collected from a carbon trading mechanism that gives more affluent countries carbon credits that they can offset against their emissions targets, if they agree to invest in projects for clean energy in the less developed countries.

Technology Transfer/Taking the Next Step

A key concern of developing countries was whether they could count on technical assistance from the industrialized countries in reducing greenhouse gas emissions. This may, for the first time, include carbon capture and storage in underground geologic formations. At the Bali conference, agreement in principle was made to find ways to make technology available to reduce greenhouse gas emission.

A key accomplishment of the conference was to have established commitments from both industrialized and developing countries to work cooperatively to solve a common problem with details to be worked out at the meeting in Copenhagen in late 2009. The new plan is intended to take effect after 2012.

> Individuals can support regional and national efforts to capture carbon dioxide and develop non-greenhouse gas–producing fuels. For many people, this may mean simply being willing to pay more for cleaner electricity and supporting government efforts to make that happen.

Emissions Trading—Cap and Trade

Emissions trading is an approach that governments use to reduce pollutants, including greenhouse gases, to certain target levels. Incentives are provided to companies or organizations to reduce the release of these gases. The government sets a cap or limit on the amount of the greenhouse gas that can be released. If a company operates below the established cap, it has a credit that it can then trade or sell to another company that is having greater difficulties meeting the cap. Enforcement of the plan often involves penalties for companies that do not meet their caps and benefits for those that do. The intent of emissions trading is to provide the greatest flexibility for companies to reach overall emissions targets with minimal impact on business. It encourages and rewards a greater contribution from organizations most able to implement changes.

A total of 27 countries in the European Union are working toward meeting their Kyoto Protocol commitments through the use of a carbon trading system. The treaty was signed in 1997 and went into effect in 2005. Under the treaty, ratifying nations that emit less than their assigned quota of greenhouse gases are able to sell credits to other countries that emit more greenhouse gases than their cap. The challenge is enforcement of the caps and verifying actual emissions, which has a cost impact.

The United States implemented an emissions trading system to enable industries to comply with the 1990 Clean Air Act, which was written to limit sulfur dioxide, a pollutant that causes acid rain. The program is intended to reduce sulfur dioxide emissions by 50 percent by the end of this decade. Regional agreements, such as by the Western Governors' Association, are currently being established to set up a cap and trade approach to carbon dioxide emissions.

The cap and trade approach has its supporters and detractors. Supporters see it as a reasonable balance that achieves emissions goals and minimizes the impact on business. Critics see it as too difficult to enforce and that efforts to track carbon credits will be prone to procrastination and abuse.

David Crane, the CEO of NRG, Inc said that "coal-fired generation is very profitable and part of that is obviously because carbon emissions from coal are still free. You can emit them in the atmosphere with no cost."[5] If releasing greenhouse gases to the atmosphere is no longer free but has a cost associated with it, alternative forms of energy will rapidly become more competitive. According to the IPCC, it may be necessary to increase the cost to emit a ton of carbon dioxide (or equivalent, $CO_{2,eq}$) to between $20 and $80 (US currency) in order to stabilize carbon dioxide levels at around 550 ppm by 2100.

[5]Interview with Jim Cramer CNBC November 7, 2007.

Turing Your Money "Green"

Some people are concerned that the response to global warming may become an impediment to economic progress. There will be costs associated with reducing greenhouse gas emissions, but there are also costs that could result from inaction.

The response to the problems of global warming and climate change can present new economic opportunities. Here are some of them:

- Developing next-generation (plug-in hybrid) cars
- Developing components for those cars (especially higher-energy-density batteries)
- Improving mass transit
- Producing and installing energy-conservation products (for new houses and to retrofit existing houses)
- Optimizing appliance efficiency
- Design and operation of low-carbon coal power (coal companies may benefit from *increased* use of coal)
- Design and operation of carbon storage (oil, coal, and oil infrastructure companies may play a role in developing sequestration facilities)
- Scientific, administrative, and legal processing of emissions trading programs
- Development and implementation of grid-connected wind and solar electricity generation

Adaptation—Global Band-Aids

Regardless of the level at which the world stabilizes greenhouse gas emissions, it is inevitable, according to the IPCC, that there will be additional warming. The present load of human-added greenhouse gases has created a commitment to at least some amount of continued warming. The earth's climate system has a lot of inertia. Where climate change cannot be avoided, adaptation may become necessary.

Two examples of adaptation include:

Venice. The mean sea level has risen 7.5 cm (3 inches) since 1897 which combined with a sinking of the land masses has increased the incidence of flooding in that city. In 1990, flood water spread across St. Marks Square roughly 7 times each year. Now the flooding occurs nearly 100 times each year threatening famous architectural landmarks. As the sea level rises, the city has become more vulnerable to flooding from storm surges. A seawall built in the fourteenth century to protect Venice is now routinely breached. Currently Venice is constructing a series of 79 huge hinged gates to separate Venice from the Adriatic Sea and protect it against storm surges.

Thames River Barrier. A set of mobile barriers were erected in the Thames river to prevent the recurrence of devastating flooding in 1953. A 3.2 m (10.5 ft) storm surge flooded parts of the UK and caused more than 300 deaths. From 1983 to 1995 there were on average1.2 closures per year. From 1996 to 2007, as a result of higher sea level, there were 6.5 closures per year.

Methods of *adaptation* will depend on the severity of the climate changes and the actual conditions in the area. Some adaptation methods that may need to be implemented include

- Constructing new levees and extending existing levees
- Changing patterns of land use, such as restricting use of areas that may become prone to flooding
- Replacing crops with those better suited to new climate conditions
- Abandoning farmland in regions subject to prolonged droughts or flooding
- Developing crop varieties with greater drought tolerance
- Increased irrigation in areas subject to droughts
- Increasing rainwater storage where periods of flooding and drought are factors
- Increasing the capacity of storm water systems
- Providing alternative habitats for the most threatened species
- Building concrete dams for glacial lakes in danger of bursting (which also may provide hydroelectric power)
- Adaptation of the agricultural marketplace in regions where increased crop yields are anticipated as a result of global warming (a positive consequence)

So Crazy It Just Might Work

A number of innovative approaches to solving the problems of global warming have been kicked around. What they may lack in immediate practicality is made up for by their creativity. Since it is difficult to anticipate the interactions that may be set into motion by changing a basic natural process, it might make sense to think carefully about any unintended consequences. Here are few of these ideas:

1. *Using phytoplankton as a carbon sink*. The oceans of the world absorb a huge amount of carbon dioxide, including a significant part of that

contributed by humans. Phytoplankton, a microscopic aquatic plant found in the oceans, consumes carbon dioxide during photosynthesis. Some scientists are exploring the possibility of enhancing the growth of these plants by using nutrients such as iron. In an experiment performed off the coast of New Zealand, researchers released 8 tons of metal in an area 5 miles (8 km) across to enhance the growth of plankton. The result was a sixfold increase in the amount of plankton and a measurable local decrease of carbon dioxide in the atmosphere. This process would have to be continued on a recurring basis. To be effective over the long term, the carbon dioxide that is removed from the atmosphere would have to be kept in the oceanic ecosystem permanently either in the phytoplankton or in organisms higher in the aquatic food chain. Since feeding plankton might benefit commercial fisheries, some commercial interest in this plan has begun to develop. However, any loss of carbon dioxide back to the atmosphere would undo the benefits of this concept. Recent tests (*Science Daily,* April 10, 2003; ScienceDaily.com) using radioactive isotope tracers show that the carbon absorbed by the plankton remained near the surface rather than descending permanently to the ocean depths. Some scientists believe that a similar process may have occurred naturally. Iron-bearing compounds may have been transported to the oceans in the past, where they contributed to promoting cold periods in the earth's past. Many scientists are very cautious about apparent solutions whose overall long-term impacts on the world's ecosystems are not fully known[6].

2. *Atmospheric reflection.* Particles in the air have an overall cooling effect on climate. Current thinking is analogous to opening up a (figurative) umbrella in the atmosphere. Such an approach would be most effective if particles were injected into the strosphere as a kind of "synthetic volcano." Several innovative (if not immediately practical ideas) include increasing the aerosol component of air pollution, seeding clouds, and placing large, lightweight (and obviously, for now, prohibitively expensive) reflective structures in orbit above the earth.
3. *Weather control.* Russian and American scientists have attempted to control the weather in the past, for example, by seeding clouds with chemicals to produce rain when and where it was needed. A new method under development involves replicating the urban heat island effect, where cities are slightly hotter than the countryside because they are darker and absorb more heat. Modifying local land reflectivity conceivably could create

[6]Fred Pearce, *Global Warming: A Beginners Guide to Our Changing Climate* New York, NY, DK Essential Science, 2002.

Figure 8-11 Algae growth pond fed by captured carbon dioxide.

almost twice as much rain 20–40 miles downwind from cities compared with upwind.

4. *Enhanced algae growth.* Carbon dioxide captured from flue gases could be used to accelerate the growth of algae. The algae then could be a source of biomass fuels such as ethanol, biodiesel, or methane. A pilot project is underway in Hawaii (Figure 8-11).

5. *Biomass fuels in power plants.* Advocates of a solution to global warming have not enthusiastically embraced biomass fuels. Much of the problem lies in the fact that, like any fuel that contains carbon in its chemical formula, ethanol (corn or cellulose) and the vegetable oils that make up biodiesel produce carbon dioxide when burned. James Hansen[7] proposed using biofuels in a power plant configured with carbon capture and storage capabilities During their growth the biofuels would remove carbon dioxide from the air. When they are burned the carbon dioxide would be captured and stored rather than released. Used in this manner, biofuels would produce needed energy and at the same time draw down carbon dioxide from the atmosphere and store it.

---

[7] *James Hansen. How Can We Avert Dangerous Climate Change? Paper based on testimony to Select Committee on Energy Independence and Global Warming, U.S. House of Representatives, 26 April 2007, posted /www.columbia.edu/~jeh1/canweavert.pdf.*

LIVE EARTH PLEDGE

On July 7, 2007, Al Gore and others, to raise awareness about global warming, organized a series of telecasts that were seen by millions of people. The program was called *Live Earth 2007.* Steps toward reducing global warming were presented in terms of the following pledge that viewers were asked to support:

- To demand that my country join an international treaty within the next 2 years that cuts global warming pollution by 90 percent in developed countries and by more than half worldwide in time for the next generation to inherit a healthy earth;
- To take personal action to help solve the climate crisis by reducing my own CO_2 pollution as much as I can and offsetting the rest to become "carbon neutral";
- To fight for a moratorium on the construction of any new generating facility that burns coal without the capacity to safely trap and store the CO_2;
- To work for a dramatic increase in the energy efficiency of my home, workplace, school, place of worship, and means of transportation;
- To fight for laws and policies that expand the use of renewable energy sources and reduce dependence on oil and coal;
- To plant new trees and to join with others in preserving and protecting forests; and
- To buy from businesses and support leaders who share my commitment to solving the climate crisis and building a sustainable, just, and prosperous world for the twenty-first century.

What You Can Do—Individual Actions

1. *Purchase and operate a fuel efficient car.*
   - Drive a car that gets at least 32 mpg.
   - Upgrade to a hybrid that gets 60 mpg.
   - Drive a plug-in hybrid to get close to 100 mpg as soon as they become available.
2. *Improve driving efficiency.*
   - Carpool when possible.
   - Reduce commute distance as much as possible.

- Maintain your car with proper tire inflation and a tuned engine.
- Avoid unnecessary trips. Call, when you can, instead of driving.
- Reduce idling. (In Japan, many drivers turn off their engines while waiting at traffic lights.) Reducing 10 minutes of idling each day can save 550 pounds of carbon dioxide per year.)
- Avoid traffic whenever possible. Traffic is inherently inefficient.

3. *Choose clean electricity.*
    - Support efforts to reduce greenhouse gas emissions by your electrical power company. If your electrical power company gives you the choice of various energy plans, choose a plan that produces the lowest amount of greenhouse gases. This may include absorbing the cost of carbon credits that support clean electricity at other sites.
    - Support any efforts on the part of local coal-fired electrical power plants to capture and contain carbon dioxide emissions. Support may be in the form of higher bills and allowing sequestration facilities in "your backyard."
    - Support local initiatives to promote nonpolluting energy on local buildings such as schools and offices. (Again, support may mean helping to pay.)
    - Install a rooftop photovoltaic system, a solar hot-water heater, or a passive solar component in your home.
4. *Reduce your consumption of energy.*
    - Turn your thermostat up in summer and down in winter. This can be done manually or by using a programmable thermostat and can reduce energy consumption while you sleep or are out. Use of a ceiling fan makes it easier to be comfortable with a higher temperature setting in the summer and sweaters in the winter.
    - Make sure that your house is buttoned up in terms of being properly insulated and weather-stripped. Keep windows and doors closed when you are using energy to heat or cool your home.
    - Use energy-efficient Energy Star appliances. A good place to start is with the refrigerator, which often is the single largest user of electrical energy in homes.
    - Use the most efficient light bulbs with the lowest required wattage. Replace incandescent lights with compact fluorescent lights. Turns lights off when they are not needed.
    - Put your computer in an energy-saving mode when you are not using it, especially one that shuts down the monitor when it is not in active use.

- Watch out for the phantom wasted standby power described earlier in this chapter. Unplug battery chargers (which use electrical power when plugged in even if nothing is connected to them) and use switchable power strips to power down televisions and other electronic devices.
- Set hot-water heater to 120°F (35°C) or below. Take shorter showers with flow-restricted shower heads to minimize hot-water use. Make sure that the hot-water heater and pipes leading from it are insulated.
- When doing laundry, use cold water instead of warm and warm water instead of hot whenever possible.
- Reduce and recycle home waste. Use minimal and recyclable packaging. Use canvas totes instead of paper or plastic grocery bags. The average home in the United States uses an estimated 1500 bags, which consume both trees and petroleum.

Final Thoughts

Throughout this book we have tried to maintain a global perspective because, after all, it is the entire Earth—the planet—that is undergoing changes. Looking down on the earth from space is probably the most powerful way to gain this perspective. Figure 8-12 shows a planet that is hungry for energy.

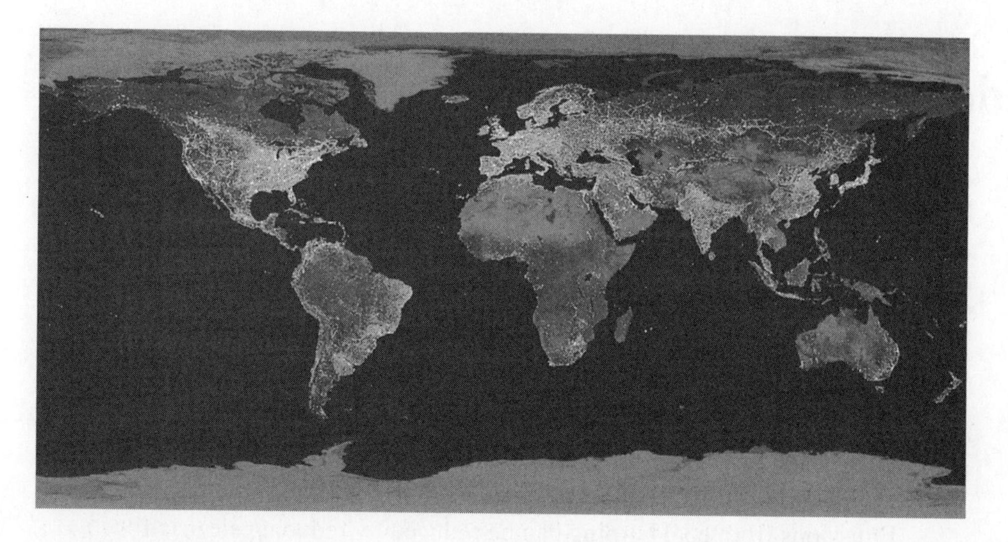

Figure 8-12 Lights from populated areas. (*Source: NASA, DMSP.*)

Figure 8-13 Like a spaceship, the earth has a built-in thermal control system that we have just begun to learn how to use. (*Source: NASA.*)

The earth's population is growing, and the earth's demand for energy is growing at three times the population growth rate. The challenge is to find a better way to provide this energy (Figure 8-13).

Key Ideas

- Greenhouse gas emissions either can continue to increase, can be held steady at a particular level, or can be reduced to a lower level.
- Some consequences of the presence of greenhouse gases in the atmosphere may be unavoidable.
- Adaptation to climate change may require providing alternate sources of water and improving flood-control provisions such as levees.
- The two main opportunities for reducing greenhouse gases are (1) reducing the amount of coal used to generate electricity and (2) reducing the amount of petroleum burned in the transportation system.
- Emissions from coal burning can be reduced by reducing electricity demand through conservation practices.

- Emissions from coal can be reduced in the short term by using more efficient IGCC power plants.
- Emissions from coal can be captured and stored in depleted oil fields, coal mines, or the oceans.
- Nonpolluting alternatives to coal-fired electricity generation include hydroelectric, wind, solar, and nuclear power.
- Although the sources of biomass fuels such as ethanol, cellulosic ethanol, and biodiesel remove carbon dioxide from the air as they grow, that carbon dioxide is released to the atmosphere when the fuel is burned. Under some, but not all, applications, the overall emissions from biomass fuels may be less when considering the overall life cycle of the fuel.
- Diesel engines are more efficient and produce fewer greenhouse gas emissions than conventional spark-ignited internal combustion engines.
- Emissions from cars can be reduced by more efficient designs, including gas–electric hybrids, plug-in hybrids, and hydrogen combustion or hydrogen fuel cell engines. The carbon dioxide reductions will be significant only if the source of electricity used emits a minimal amount of carbon dioxide.

Review Questions

1. On average, how often is a new large coal-fired electricity power plant currently coming online in China?
   (a) Each week
   (b) Each month
   (c) Each quarter
   (d) Each year
2. Which of the following is an example of carbon sequestration?
   (a) Turning coal to more efficient syngas
   (b) Replacing coal-fired plants with alternative sources of energy
   (c) Capturing and storing the carbon dioxide from a coal-fired plant and sending it to an abandoned coal mine
   (d) Using natural gas instead of coal
3. Which of the following does *not* produce greenhouse gases?
   (a) Nuclear power
   (b) Ethanol

(c) Biodiesel

(d) Natural gas

4. Which of the following required a Supreme Court decision recently in the United States to be officially considered a pollutant?

   (a) Halocarbons

   (b) Carbon monoxide

   (c) Ozone

   (d) Carbon dioxide

5. What percent of the world's electricity is currently generated by technology that does not produce greenhouse gas emissions?

   (a) 10 percent

   (b) 30 percent

   (c) 50 percent

   (d) 75 percent

6. Which of the following would be found in the tailpipe of a vehicle that burns hydrogen?

   (a) Oxygen

   (b) Water vapor

   (c) Carbon dioxide

   (d) Ozone

7. What is a primary advantage of the IGCC approach to coal-generated electricity?

   (a) Less coal is consumed.

   (b) It is easier to remove carbon dioxide that is produced.

   (c) Costs are less than those of standard coal plants without capture and storage.

   (d) Less methane is given off.

8. Which of the following represents adaptation to climate change?

   (a) Building water reservoirs and pipelines to villages in the foothills of the Himalayas

   (b) Driving hybrid cars

   (c) Building nuclear power plants

   (d) Turning thermostats lower

9. In an emissions trading arrangement, what is a company or country that is unable to meet its greenhouse gas emissions target obligated to do?
   (a) Shut down its operations
   (b) Meet the emissions targets within a set time period
   (c) Pay an organization that can reduce emissions beyond that organization's requirements
   (d) Make its best good faith effort to meet its commitment in the future
10. Which of the following is currently best able to compete economically with coal-generated electricity?
    (a) Wind
    (b) Photovoltaics
    (c) Nuclear
    (d) IGCC with capture and storage
11. Which of the following characterizes most of the new photovoltaic systems currently being installed?
    (a) Roadside signs
    (b) Grid-connected
    (c) Remote power
    (d) Irrigation pumping stations
12. What commitment is required for countries ratifying the Kyoto Protocol?
    (a) Zero carbon dioxide emissions by 2035
    (b) To do their best on a voluntary basis
    (c) Eliminate chlorofluorocarbons (CFCs) and other gases that attack stratospheric ozone
    (d) Reduce greenhouse gas emissions to agreed-on levels

Final Exam

1. Why are anomalies often preferred to absolute temperature measurements?
   (a) They are easier to measure.
   (b) They require less computer memory.
   (c) They reduce the likelihood of calibration errors.
   (d) They use special thermometers.
2. What does a *paleoclimatologist* study?
   (a) Historical climate records
   (b) Ocean salinity
   (c) Satellite data
   (d) Tropical storm intensity
3. Which of the following is the most significant cause of increased temperature in cities?
   (a) Global warming
   (b) Heat island effect

(c) Ozone
(d) Waste heat from nuclear power plants

4. How much does the heat island effect contribute to global warming?
   (a) Much less than 1 percent
   (b) 2 percent
   (c) 5 percent
   (d) 10 percent
5. Why is the stratosphere cooling?
   (a) The sun is getting less intense
   (b) Less ozone results in slightly less absorption of incoming solar ultraviolet energy
   (c) The earth is becoming more reflective
   (d) Greater evaporation cools the stratosphere
6. How do the TOPEX satellites measure sea level?
   (a) Visual imagery
   (b) Infrared imagery
   (c) Ultraviolet sensors
   (d) Reflected radar waves
7. What trend is noticeable in the Palmer Drought Severity Index?
   (a) There is no statistical trend.
   (b) Drought conditions are increasing worldwide.
   (c) Drought conditions are decreasing worldwide.
   (d) Cycles are coinciding with natural climate oscillations.
8. How has relative humidity been changing in recent years?
   (a) No significant change
   (b) Increasing
   (c) Decreasing
   (d) Varying with local temperature
9. How has the amount of atmospheric water vapor been changing in recent years?
   (a) No significant change
   (b) Increasing

(c) Decreasing

(d) Varying with local temperature

10. Besides melting of glaciers and sea ice, what else causes global sea level rise?

(a) Increased precipitation

(b) Permafrost melting

(c) Arctic sea ice melting

(d) Thermal expansion

11. Why is it difficult to be sure whether the number of category 4 and 5 hurricanes are actually becoming more frequent?

(a) Modern structures are better able to withstand storm damage.

(b) Better surveillance recently may have improved detection of intense conditions.

(c) Precise proxy records do not exist for wind speed.

(d) Present day cities may impede the intensity of wind conditions.

12. About how much of the earth's surface is covered by permafrost?

(a) 20–25 percent

(b) 8 percent

(c) 3 percent

(d) Less than 1 percent

13. Why does Great Britain have a milder climate than other places at the same latitude?

(a) Fewer clouds

(b) More direct sun

(c) Heat distributed by the thermohaline circulation (THC).

(d) Greater sources of geothermal heat

14. Which scientist first proposed the idea that planets reradiate heat received from the sun back into space—an idea which became known as the greenhouse effect?

(a) John Tyndall

(b) Joseph Fourier

(c) Svante Arrhenius

(d) Guy Stewart Callendar

15. By about how much does the solar "constant" vary during the sun's 11 year sunspot cycle?
    (a) 10 percent
    (b) 7 percent
    (c) 1–2 percent
    (d) 0.1 percent
16. By how much does the solar intensity received by the earth vary during the earth's annual revolution about the sun?
    (a) 10 percent
    (b) 7 percent
    (c) 1–2 percent
    (d) 0.1 percent
17. What is the cause for the Milkanovich cycles?
    (a) Gravitational pull of Jupiter and Saturn
    (b) Solar intensity variations
    (c) Solar wind variations
    (d) Volcanic activity
18. If nearly 90 percent of the energy received from the sun goes into the oceans, why has the average temperature of the oceans risen by what may seem like a small amount (only 0.1°C) over the past few years?
    (a) Heat absorbed by the oceans is quickly transferred by convection to the atmosphere.
    (b) The oceans have a very large heat capacity and require a large amount of heat to result in a measurable temperature change.
    (c) Melting glaciers and sea ice keep the oceans at a lower temperature.
    (d) The heat from the ocean is reradiated back into space.
19. How is solar output related to sunspot cycles?
    (a) Solar output is unchanged throughout the sunspot cycle.
    (b) Solar output is slightly lower when there are fewer sunspots.
    (c) Solar output is slightly higher when there are fewer sunspots.
    (d) Solar output increases and decreases every 5 ½ years.
20. What happens to the temperature of melting ice or snow as it melts?
    (a) It remains constant at several degrees above 0°C.
    (b) The more heat it absorbs, the hotter it gets.

(c) It remains constant at several degrees below 0°C (32°F).

(d) It remains constant at close to 0°C (32°F).

21. Which of the following has the highest specific heat? (In other words, which of the following require the maximum heat for the temperature to change a given amount?)

(a) Air

(b) Water

(c) Rock

(d) Carbon dioxide

22. Why did the Mount St. Helens eruption have less of an impact on global climate than the Mt. Pinatubo eruption?

(a) Mount St. Helens emitted less ash during the eruption.

(b) Mount St. Helens erupted during La Niña conditions.

(c) Mount St. Helens released ash mainly into the troposphere where it was washed away by precipitation.

(d) Mount St. Helens released ash mainly into the stratosphere where it was quickly diluted.

23. When do airline contrails have the greatest warming impact on atmospheric temperature?

(a) Clear nights

(b) Clear mornings

(c) Clear afternoons

(d) Rainy days

24. What is the effect of melting ice on the reflectance (albedo) of the earth?

(a) No effect

(b) Reflectance increases

(c) Reflectance decreases

(d) Reflectance increases over water but decreases on land

25. Why is Venus warmer than Mercury despite the fact that it is further from the sun?

(a) Mercury has more clouds.

(b) Carbon dioxide on Venus causes a greenhouse effect.

(c) Water on Venus absorbs more heat.

(d) The surface of mercury has a higher specific heat than Venus.

26. What is meant by an *enhanced* greenhouse effect?
    (a) Any absorption of infrared energy by greenhouse gases
    (b) Absorption of ultraviolet energy by greenhouse gases
    (c) Absorption of infrared energy by human-generated gases
    (d) Absorption of infrared energy by naturally produced gas levels
27. In what form is carbon dioxide primarily stored in the ocean?
    (a) As carbonic acid
    (b) As small bubbles in the water
    (c) In plants such as water
    (d) In corals
28. How does the integrated combined cycle (IGCC) make it easier to remove carbon dioxide from coal?
    (a) Carbon dioxide is returned to a solid form
    (b) Carbon dioxide is less diluted by nitrogen
    (c) No carbon dioxide is produced
    (d) Carbon dioxide is converted to hydrogen
29. Which fuel has the highest mass energy density (energy produced for a given *mass* of fuel burned)?
    (a) Coal
    (b) Oil
    (c) Natural gas
    (d) Hydrogen
30. Which fuel has the lowest volume energy density (energy produced for a given *volume* of fuel burned)?
    (a) Coal
    (b) Oil
    (c) Natural gas
    (d) Hydrogen
31. Which fuel produces the most carbon dioxide for a given *mass* of fuel burned?
    (a) Coal
    (b) Oil
    (c) Natural gas
    (d) Hydrogen

32. What happens to the concentration of O-18 in precipitation as the atmospheric temperature goes up?
    (a) O-18 concentration changes but does not depend on the atmospheric temperature.
    (b) O-18 concentrations remain constant regardless of the temperature.
    (c) There is more O-18 as the temperature becomes warmer.
    (d) There is less O-18 as the temperature becomes warmer.
33. Which of the following greenhouse gases absorb infrared light most strongly for a given concentration of the gas?
    (a) Nitrogen (N_2)
    (b) Oxygen (O_2)
    (c) Methane (CH_4)
    (d) Carbon dioxide (CO_2)
34. What is an isotope?
    (a) Atoms with differing number of neutrons
    (b) Radioactive atoms
    (c) Charged atoms
    (d) Atoms that react with uranium
35. The half-life of carbon-14 is 5730 years. If an ice core sample has 25 percent of the carbon-14 than found in the atmosphere, how old is the sample?
    (a) 2,865 years
    (b) 5,730 years
    (c) 11,400 years
    (d) 17,190 years
36. How can global warming increase the risk of malaria?
    (a) Introducing new areas of standing water through flooding and precipitation
    (b) Increasing ocean pH causing local migration of mosquito populations
    (c) Spreading drought conditions destroying the mosquitoes' habitat
    (d) Increasing the likelihood of statistically warmer nights
37. What obstacles prevent the more widespread use of hydrogen as a fuel?
    (a) Developing an energy-efficient way to produce hydrogen
    (b) Establishing an effective way to distribute hydrogen

(c) Finding methods for storing hydrogen in a reduced volume

(d) All of the above

38. Which form of solar energy is the fastest growing currently?

(a) Charging batteries for overnight home needs

(b) Roadside signs

(c) Electricity generation in excess of a home's requirements distributed to the utility's electrical power grid

(d) Decomposing sea water to produce hydrogen fuel

39. Which of the following is being considered as a possible form for long-term storage of carbon dioxide near the ocean floor?

(a) Soluble calcium bicarbonate

(b) Solid calcium carbonate

(c) Carbon dioxide bubbles

(d) Absorption by aquatic plants

40. If halocarbons are such strong absorbers of infrared energy, why do they not have a more significant role in warming the atmosphere?

(a) They last in the atmosphere only for a very short time.

(b) They absorb wavelengths that are already being strongly absorbed by other greenhouse gases.

(c) They are present in relatively small concentration.

(d) They absorb only a very small portion of the infrared spectrum.

41. What gas is ozone produced from?

(a) Oxygen (O_2)

(b) Methane (CH_4)

(c) Carbon dioxide (CO_2)

(d) Carbon tetrachloride (CH_4)

42. How concentrated is a gas that is 382 ppm?

(a) 382 molecules of the gas for every billion molecules of the atmosphere

(b) 382,000,000 molecules of the gas for every million molecules of the atmosphere

(c) 38.2 percent of the gas in the atmosphere

(d) 382 molecules of the gas for every million molecules of the atmosphere

43. What does $GtCO_2$,eq measure?
    (a) $GtCO_2$,eq includes just human generated carbon dioxide.
    (b) $GtCO_2$,eq includes the mix of all greenhouse gases based on their contribution to infrared absorption compared to carbon dioxide
    (c) $GtCO_2$,eq includes only naturally produced carbon dioxide
    (d) $GtCO_2$,eq includes all greenhouse gases except for carbon dioxide
44. How do "carbon emissions" as measured by GtC (Giga-tons of carbon) compare with carbon dioxide emissions, $GtCO_2$ (Giga-tons of carbon dioxide)?
    (a) They are basically the same.
    (b) GtC includes emissions from fossil fuels only.
    (c) GtC is 27.3 percent smaller than $GtCO_2$ reflecting the percentage of carbon in carbon dioxide by weight.
    (d) GtC is larger because solid carbon has a greater density than carbon dioxide which is a gas.
45. How do most of the emissions of the greenhouse gas, nitrous oxide, get into the atmosphere?
    (a) Use of nitrogen based fertilizers
    (b) From the dental industry
    (c) From combustion of coal
    (d) From nuclear power
46. What was the intent of the 1989 Montreal Protocol?
    (a) Limiting greenhouse gas emissions
    (b) Phasing out chlorofluorocarbons (CFC's)
    (c) Requiring more alternative energy sources
    (d) Energy conservation
47. How do we know that carbon dioxide in the atmosphere comes from burning fossil fuels, rather than as a natural by-product of ocean warming?
    (a) Atmospheric oxygen levels have increased.
    (b) The southern hemisphere produces more greenhouse gases than the northern hemisphere.
    (c) Carbon-14 found in atmospheric carbon dioxide is not present in fossil fuels.
    (d) Ocean pH has been increasing.

48. Which of the following supports the idea that elevated levels of carbon dioxide in the atmosphere comes from human sources?
    (a) Carbon dioxide levels are highest near industrial sources especially in the northern hemisphere.
    (b) Only glaciers in industrial countries are melting.
    (c) The hole in the ozone over Antarctica is becoming smaller.
    (d) Increasing methane levels in the air come from combustion of coal.
49. Which properties of greenhouse gases are incorporated into the index called the GWP (Global Warming Potential)?
    (a) How likely it is to be released into the atmosphere
    (b) Cost to capture and ease of removal from smokestacks
    (c) Concentration of the gas compared to preindustrial levels
    (d) Infrared absorption and lifetime in the atmosphere
50. Which of the following has the highest GWP?
    (a) Carbon dioxide (CO_2)
    (b) Methane (CH_4)
    (c) A hydrofluorocarbon such as CH_3CF_3
    (d) Nitrous oxide (N_2O)
51. What is radiative forcing?
    (a) Release of radioactive waste from a nuclear reactor
    (b) An increase or decrease in the amount of energy received by the earth
    (c) Mandatory carbon tax imposed on countries that exceed target emission levels
    (d) Forcing carbon dioxide and other greenhouse gases out of the atmosphere
52. Which of the following is the most significant example of a negative radiative forcing?
    (a) Aerosols
    (b) Greenhouse gases
    (c) Melting snow
    (d) Destruction of stratospheric ozone

53. What is the one way that climate change can increase the spread of diseases such as hemorrhagic dengue fever?
    (a) Higher greenhouse gas concentrations increase mosquito reproduction.
    (b) Higher temperatures result in spread of bacteria.
    (c) Migration of mosquito populations to higher latitude regions increase the exposure of a greater human population density that is less prepared to cope with tropical diseases.
    (d) Lower ultraviolet radiations at the earth's surface result in fewer mosquitoes being killed.
54. What is the single largest source of electrical power generation on the earth?
    (a) Oil
    (b) Natural gas
    (c) Coal
    (d) Hydroelectric
55. What current industrial practice may most readily be adapted into a method for carbon sequestration?
    (a) Use of carbon dioxide to extract oil from nearly depleted fields
    (b) Producing cement from quicklime, $CaCO_3$
    (c) Burning ethanol in internal combustion engines
    (d) Hydroelectric power generation
56. What is an example of chemical sequestration?
    (a) Storing carbon dioxide underground in sites such as depleted oil and gas fields, coal seams, and brine fields
    (b) Fixing the carbon dioxide in biomass such as algae
    (c) Burning coal using IGCC (Integrated gasification combined cycle) technology
    (d) Forming stable minerals such as calcite ($CaCO_3$) or magnesite ($MgCO_3$)
57. Why is it difficult to remove carbon dioxide from the smokestacks of coal fired power plants?
    (a) The gases are too hot.
    (b) A large amount of atmospheric nitrogen is in the waste stream.

(c) The gases are explosive.

(d) The gases do not dissolve in any known solvent.

58. How is the use of capture and storage expected to affect the overall amount of coal that is needed by a power plant?

(a) Increases

(b) Decreases

(c) Requires the same amount but only the purest coal

(d) No effect

59. Which of the following potentially requires the least amount of electrical energy to produce a given amount of light?

(a) Incandescent bulbs

(b) Fluorescent bulbs

(c) Light emitting diodes (LED's)

(d) Halogen light bulbs

60. Why is *an inverter* used for grid-connected photovoltaic systems?

(a) To distribute direct current (DC) power for use at the site where it is generated

(b) To convert DC power from the solar panels into alternating current (AC)

(c) To charge batteries

(d) To store power during periods of cloudiness or nighttime hours

61. Presently, what material do most commercial solar cells use as the base material that absorbs light energy?

(a) Gallium arsenide

(b) Gallium nitride

(c) Cadmium telluride

(d) Silicon

62. What is the most widespread use of solar power around the world?

(a) Solar hot water heating

(b) Solar space heating

(c) Solar electrical generation

(d) Solar direct heat to produce steam for electrical generation

63. China leads the world in installations of solar heating panels. What feature do many of the Chinese solar heating panels use?
    (a) Flat-plate designs
    (b) Evacuated tubes
    (c) Combined heat and photovoltaic panels
    (d) Large centralized thermal heating system
64. What is a significant additional cost that limits the use of solar hot water heating in many cold climates?
    (a) The need for special anti-reflective coatings
    (b) Avoiding reflection from snow
    (c) The need for freeze protection
    (d) Providing sufficiently large storage
65. Most geothermal power plants planned for construction are binary-cycle plants. What is a binary-cycle geothermal system?
    (a) Water in the geothermal reservoir is passed through a heat exchanger that transfers the heat to a separate pipe containing fluids with a much lower boiling point
    (b) Steam from underground geysers directly turn electrical turbines.
    (c) Geothermal energy is combined with a coal fired plant to improve its efficiency.
    (d) Waste steam from the geothermal plant is reused a second time.
66. Geothermal energy can be used for directly heating or cooling homes and for reducing the requirement from other energy sources. What is the temperature of the earth in many places below the surface?
    (a) 35°C (95°F)
    (b) 13°C (55°F)
    (c) 1.5°C (35°F)
    (d) –19°C (–2°F)
67. In the United States, how does the average gasoline consumption of 1.3 gallons per day compare with the average for the rest of the world?
    (a) Ten times
    (b) About the same

(c) About in the middle

(d) Twice

68. What enables the gasoline engine in a hybrid gasoline-electric vehicle to be much less massive and, as a result, much more efficient at cruising speeds?

(a) It uses more efficient fuels.

(b) The hybrid vehicle does not accelerate well.

(c) An electric motor contributes power during acceleration.

(d) The motor runs like a diesel engine.

69. What is necessary for automobile designs that use pre-charged batteries (such as plug-in hybrids) to contribute to a reduction of greenhouse gas emissions?

(a) Highways must be more leveled to reduce the amount of regenerative braking that takes place.

(b) The source of electricity used to pre-charge batteries must generate a minimal amount of greenhouse gases.

(c) Lithium ion batteries are needed.

(d) Hybrids should use biomass fuels to reduce overall emissions.

70. Burning ethanol reduces global gas emissions . . .

(a) if the ethanol comes from corn.

(b) if automobile engines are better adapted to burn ethanol so that the combustion is more efficient.

(c) if ethanol is used in hybrid cars.

(d) if more carbon dioxide is removed from the atmosphere during the entire ethanol growth and distribution cycle than is released during combustion.

71. What is considered by scientists to be an attainable level of CO_2eq to stabilize at to avoid the most severe consequences of global warming?

(a) Ten percent below preindustrial levels

(b) Greenhouse gas concentrations returned to preindustrial levels

(c) Roughly double preindustrial levels

(d) Five times preindustrial levels

72. What is the scope of actions needed worldwide to stabilize greenhouse gas emissions?

(a) Mileage for 2 billion cars increased from 30 to 60 mpg

(b) Two actions equivalent to increasing 2 billion cars from 30 to 60 mpg

(c) Seven actions equivalent to increasing 2 billion cars from 30 to 60 mpg

(d) Twelve actions equivalent to increasing 2 billion cars from 30 to 60 mpg

73. How can biomass fuel in power plants be made to reduce greenhouse gas emissions?

(a) Replacing productive farmland with acreage dedicated to ethanol fuel production

(b) Combining with carbon capture and storage

(c) Combining with high-octane petroleum distillates

(d) Combining with high-purity coal

74. About how much more expensive is coal-generated electrical power expected to be if carbon dioxide emissions are captured and stored?

(a) The same

(b) 25 percent more expensive

(c) Double the cost

(d) Triple the cost

75. What is a reasonable expectation for the fuel efficiency for a near-future plug-in hybrid vehicle?

(a) 32 mpg

(b) 40 mpg

(c) 60 mpg

(d) 100 mpg

APPENDIX A

Glossary

Aerosols Airborne solid or liquid particles with a typical size between 0.01 and 10 μm that reside in the atmosphere for at least several hours. Aerosols may be of either natural or human in origin. Aerosols may influence climate either directly through scattering and absorbing radiation or indirectly by facilitating the formation of clouds.

Albedo The fraction (or percentage) of solar radiation reflected by a surface. Snow-covered surfaces have a high albedo; the albedo of soil ranges from high to low; vegetation-covered surfaces and oceans have a low albedo. The earth's albedo varies mainly through varying cloudiness and snow, ice, leaf area, and land-cover changes.

Anthropogenic Resulting from human as opposed to natural impacts on climate.

Carbon intensity A measure of carbon emissions compared with a given level of economic activity. Carbon intensity is an index of how efficiently a country (or other economic entity) is using energy to meet its needs.

Climate feedback An interaction among processes in the climate system in which a change in one process triggers a secondary process that, in turn, influences the first one. A positive feedback intensifies the change in the original process, and a negative feedback reduces it.

Climate model A numerical representation of the climate based on the physical, chemical, and biologic properties of its components and their interactions and feedback processes and accounting for all or some of its known properties.

Climate sensitivity How much the global mean surface temperature will go up following a doubling of the atmospheric carbon dioxide concentration.

Commitment The tendency of the climate to continue changing as a result of past greenhouse gas emissions even if gases in the atmosphere are held fixed at already established levels.

Cryosphere The component of the earth consisting of all snow, ice, and permafrost.

El Niño southern oscillation (ENSO) El Niño is a warm water current that flows periodically along the coast of Ecuador and Peru, disrupting the local fishery. This ocean event is associated with a fluctuation in the tropical surface pressure pattern and circulation in the Indian and Pacific Oceans called the *southern oscillation.* Together the combined effects in the atmosphere and ocean are known as the *El Niño southern oscillation* (ENSO). During an El Niño event, the prevailing trade winds weaken, and the equatorial countercurrent strengthens, causing warm surface waters in the Indonesian area to flow eastward to overlie the cold waters of the Peru current. This event has great impact on the wind, sea surface temperature, and precipitation patterns in the tropical Pacific. It has climatic effects throughout the Pacific region and in many other parts of the world.

Electromagnetic wave Energy such as light received from that sun. Electromagnetic waves include radio waves, microwaves, infrared light, visible light, ultraviolet light, x-rays, and gamma rays. Electromagnetic radiation differs in the wavelengths of the waves. For instance, infrared waves, which are emitted by the earth after its is warmed by the sun, have a longer wavelength than visible light waves received from the sun.

EOS NASA's Earth Observing System.

EPA U.S. Environmental Protection Agency.

ESA European Space Agency.

Glacial lake outburst flood A sudden and often catastrophic failure of a dam formed from glacial ice or from deposits (called terminal moraine) left by a receding glacier. When the dam fails, the lake containing a buildup of glacial meltwater is released causing local flooding. Increased melting from global warming puts greater pressure on these dams.

Global isostatic adjustment Response of the earth's crust to changes in the mass supported by land resulting from the melting or buildup of ice mass. Isostatic adjustment must be taken into account when determining sea level relative to an adjacent coastline whose level also may have changed. This is also referred to as *postglacial rebound.*

Global warming potential (GWP) A measure of the impact a greenhouse gas has compared with carbon dioxide. It takes into account how long the gas remains in the atmosphere and how effective the gas is in absorbing infrared radiation.

Greenhouse effect Absorption of thermal infrared radiation emitted from the earth's surface by gases in the atmosphere. The natural greenhouse effect is the result of gases present before industrialization and worldwide combustion of fossil fuels. The enhanced greenhouse effect is the additional absorption of infrared radiation by gases added to the atmosphere as a result of human activities.

Greenhouse gas Greenhouse gases are components of the atmosphere, both natural and anthropogenic, that absorb (and emit) radiation at specific wavelengths within the spectrum of infrared radiation emitted by the earth's surface, the atmosphere, and clouds. This property causes the greenhouse effect. Water vapor (H_2O), carbon dioxide (CO_2), nitrous oxide (N_2O), methane (CH_4), and ozone (O_3) are the primary greenhouse gases in the earth's atmosphere. There are also a number of entirely human-made greenhouse gases in the atmosphere, such as the halocarbons and other chlorine- and bromine-containing substances.

Heat island effect The increased temperature of an urban environment caused by structures, pollution, and heat-generating activities causing absorption of a greater proportion of the sun's energy than rural areas.

Infrared radiation Electromagnetic waves emitted by the earth at wavelengths that are too long to be visible. There is a small component of infrared light in the sunlight that hits the earth, but all the electromagnetic waves emitted by the earth after it is warmed by the sun are in this wavelength range. Absorption of this infrared radiation by the atmosphere is what is called the *greenhouse effect.*

Instrumental record The period of climate measurement that includes actual temperature measurements rather than indirect analytical methods used to decipher past temperatures from ice cores, tree rings, and coral layers. The instrument record generally is considered to have started in the late 1800s.

IPCC Intergovernmental Panel on Climate Change, which was established in 1988. It is open to all members of the United Nations and World meteorological Organization. The IPCC's role is to assess information on climate change. It does not carry out research, nor does it monitor climate data.

Isotope Atoms of the same element that have different numbers of neutrons. Isotopes are used to determine the age of fossil records and to determine the sources of gases released into the atmosphere.

La Niña The opposite of an El Niño event is called *La Niña*. La Niña forms part of the ENSO cycle.

Methane clathrate A semifrozen mix of methane gas and ice typically found in the northern hemisphere's tundra permafrost regions and in sediment layers on the ocean floor. Sudden release of large amounts of methane from methane clathrate deposits has been proposed as a cause of past and possibly future climate changes.

NOAA National Oceanic and Atmospheric Administration, which was established in 1970 to conduct research and gather data about the global oceans, atmosphere, space, and sun.

North Atlantic oscillation (NAO) Opposing variations of barometric pressure near Iceland and near the Azores. This affects the strength of the main westerly winds crossing the Atlantic into Europe. It is the dominant mode of winter climate variability in the North Atlantic Region, ranging from central North America to Europe.

Paleoclimatogy The study of climate prior to the use of measuring instruments, including historic and geologic time.

ppm (ppb) parts per million (parts per billion). A measure of how concentrated a gas is in the atmosphere.

Proxy data Indirect climate data other than those derived from measurement instruments. Temperature and gas concentrations obtained from ice cores, boreholes, tree rings, and coral sections are proxy data.

Radiative forcing A change in the environment that directly affects the amount of energy being absorbed by the earth. A positive forcing tends to warm the surface of the earth, and a negative forcing tends to cool the surface.

Sequestration The process of (permanently) storing carbon someplace other than the atmosphere. Biologic sequestration includes absorption of carbon dioxide by trees. Physical approaches include separation of carbon dioxide from smokestacks and storage in underground empty oil reservoirs or coal seams.

Sink Any process, activity, or mechanism that removes a greenhouse gases from the atmosphere.

Thermohaline circulation Synonymous with *meridonal overturning circulation* (MOC). This is a global ocean current that transfers massive amounts of heat from near the equator to higher latitudes in a process often compared to a conveyor belt. This circulation is driven by a combination of cold temperatures and high salinity near the ocean surface.

Troposphere The lowest part of the atmosphere from the surface to about 10 km in altitude in midlatitudes (ranging from 9 km in high latitudes to 16 km in the tropics, on average) where clouds and "weather" occur. In the troposphere, temperatures generally decrease with height.

APPENDIX B

Milestones in the History of Climate Change

115,000 years ago Interglacial warming period characterized by rising sea levels, melting of Antarctic ice sheets, receding glaciers, and inland flooding (see Chapter 2 for additional details about this period).

11,000 years ago The last ice age took place.

800–1100 The Vikings settled in Greenland at a time when it was experiencing an unusually warm climate.

1250 Start of the little ice age, which was a cold period possibly brought on by volcanic eruptions. The Arctic ice pack in Greenland moved southward, making navigation in Greenland waters increasingly hazardous and life far more difficult for the descendants of the Vikings in Greenland.

1815 Mount Tambora in what is now Indonesia erupted.

1816 Aerosols produced by the Mount Tambora eruption caused colder conditions around the world. The year had become known as "the year without a summer."

1827 Joseph Fourier recognized that gases in the atmosphere might increase the surface temperature of the earth, establishing the groundwork for what would later be known as the *greenhouse effect.* He developed the concept of planetary energy balance, in which planets receive energy from the sun and lose energy by infrared radiation.

1860s John Tyndall studied absorption of light by different gases, including coal gas, carbon dioxide, and water vapor. In addition to water vapor, Tyndall showed that carbon dioxide dissolved in water absorbs a great deal of heat energy. He saw this as possible cause of climate change and a possible explanation for the advance and retreat of glaciers.

1896 Svante Arrenius quantified the relationship between carbon dioxide in the air and global warming. His prediction was slightly higher than what is currently accepted but laid the groundwork for modeling climate sensitivity.

1912 Alfred Wegener proposed the theory of continental drift. Reorientation of continents is thought to have played a role in climate history because of the way it altered the reflectivity of the earth's surface and, as a result, its temperature.

1920 Milutin Milanković, a Serbian civil engineer and mathematician, calculated the cycles caused by changes in the earth's orbit around the sun. Variations associated with eccentricity, axial tilt, and precession of the earth's orbit resulted in 100,000-year ice age cycles over the last few million years.

1938 Guy Stewart Callendar, a British coal engineer, analyzed temperature measurements taken from weather stations and concluded that there is an increase in global temperature as carbon dioxide increases. He attributed this rise in temperature to the accumulation of carbon dioxide in the atmosphere as a result of burning fossil fuels.

1950 Hans Suess studied the distribution of carbon-14 and tritium in the oceans and atmosphere. On the basis of radiocarbon analyses of annual growth rings of trees, he contributed to the calibration of the radiocarbon dating scale and the study of the magnitude of the dilution of atmospheric radiocarbon by carbon dioxide from fossil fuels burned since the industrial revolution. This dilution is known as the *Suess effect.*

1958 Charles Keeling began carbon dioxide measurements at the Mauna Loa Observatory in the Hawaiian Islands. His work identified human-added carbon dioxide that progressively built up in the atmosphere each year.

1960s Carl Sagan, an American scientist, determined that the atmosphere of Venus is extremely hot and dense. He perceived global warming as a growing, human-induced danger similar to the transformation of Venus into a hot, uninhabitable planet caused by the presence of greenhouse gases. Sagan's predictions about the surface of Venus were confirmed by the *Mariner 2* spacecraft, whose mission he helped plan.

1970 Clean Air Act established to empower the U.S. Environmental Protection Agency (EPA) to regulate air quality in the United States. This significantly upgraded a previous enactment of the law in 1963.

1980 Mount St. Helens erupted in Washington State. The impact on climate was minimal because particulates were released primarily in the troposphere and were washed out relatively quickly by precipitation.

1982 The El Chinchon eruption in Mexico caused a decrease in global temperature. Aerosols emitted into the stratosphere by the volcano persisted for months.

1984 Mobile flood barrier completed in the Thames river intended to prevent recurrence of previous disastrous flooding.

1988 James Hansen, a NASA scientist, testified about global warming before the U.S. Congress.

1988 The Intergovernmental Panel on Climate Change (IPCC) was established by two United Nations organizations, the World Meteorological Organization (WMO) and the United Nations Environment Programme (UNEP), to assess the risk of human-induced climate change.

1989 The Montreal Protocol on substances that deplete the ozone layer was established. Control of these substances was intended to protect the ozone layer. The substances, however, are also greenhouse gases.

1990 The IPCC first assessment report was completed.

1991 The Mount Pinatubo eruption in the Philippines resulted in a decrease in global temperature.

1992 The Rio Treaty signed by over 160 countries, committing them to (voluntarily) reduce emissions of greenhouse gases.

1993 One of the most destructive floods in U.S. history occurred involving the Missouri and Mississippi River valleys.

1995 The Larsen A ice shelf in Antarctica, which had been stable for hundreds of years, disintegrated, losing 1700 km^2 in a week.

1997 The Kyoto Protocol was negotiated in Kyoto, Japan, in December. Countries ratifying this protocol committed themselves to reducing their emissions of carbon dioxide and five other greenhouse gases. Notable countries not ratifying the treaty included the United States, China, India, and Australia. The United States and Australia were the only industrialized countries that did not sign the agreement. China and India have ratified the Kyoto protocol but have not committed to specific emission targets. (Australia eventually signed on to the Kyoto protocol at the start of the follow-on conference in Bali in 2007).

1997 A severe El Niño event produced the hottest year on record.

2001 The IPCC released its third assessment report.

2002 The Larsen B ice shelf in Antarctica broke up, losing 3250 km^2 of ice 220 m. This section of the shelf may have been stable for thousands of years.

2003 A heat wave in Europe resulted in the deaths of an estimated 35,000 people.

2005 Hurricanes Katrina, Rita, and Wilma caused widespread destruction and environmental harm to coastal communities in the U.S. Gulf Coast region.

2007 The IPCC released its fourth assessment report (AR4).

2007 Record drought conditions threaten communities in the southeastern part of the United States.

2007 The Nobel Peace Prize was awarded to the IPCC and Al Gore for work on climate change.

2007 The United Nations Climate Change Conference held in Bali to reach international agreement on concrete steps to be taken in response to the IPCC's fourth assessment report (AR4). The conference extends commitments to greenhouse gas reductions established as part of the Kyoto Protocol.

APPENDIX C

Satellites That Monitor Weather and Climate

Satellite	Mission/Function
Geostationary Operational Environmental Satellite (GOES) Program	Visual image of earth from high altitude above a fixed spot on the earth. This satellite provides the TV hurricane maps. Continuously monitors severe weather conditions and tracks their movements, including cloud cover, storm patterns, tornadoes, flash floods, hurricanes, volcanic eruptions, and forest fires. Estimates rainfall, snowfall, and overall extent of snow cover and sea and lake ice.
Polar orbiting satellites [such as the Tiros (Television Infrared Observation) satellites operated by NOAA or the National Oceanic and Atmospheric Administration]	Visual image of earth from low orbit—cloud cover, storm patterns, infrared tracking.
The Defense Meteorological Satellite Program (DMSP)	Visual image of earth from low orbit; monitors meteorologic, oceanographic, and solar-terrestrial physical environments.
SOHO	Observes the sun, sunspots, and solar flares.
Ocean TOPography Experiment (TOPEX) (*Poseidon and Jason*)	Uses radar altimetry to monitor ocean wave height and flow; studies sea level rise, ENSO, and ocean currents
Gravity Recovery and Climate Experiment (GRACE)	Local gravitational field and glacial mass; consists of twin satellites flying in formation and performing precise high-resolution gravitational measurements; changes in the gravitational field directly below the spacecraft cause small changes in the orbit that are detected by state-of-the-art Global Positioning Satellite (GPS) instruments. This enables tracking of the mass of snow and ice fields to determine the extent of melting.
Upper Atmosphere Research Satellite (UARS)	Performed a wide range of measurements in the upper atmosphere focusing on ozone and chemicals involved with its synthesis or breakdown.
Cloud-Aerosol Lidar and Infrared Pathfinder Satellite Observation (CALIPSO)	Joint U.S. (NASA) and French mission (CNES) that studies how clouds and aerosols form, evolve, and affect weather and climate; CALIPSO coordinates with CloudSat and together they fly in formation with three other satellites.
EnviSat	Operated by the European Space Agency (ESA). Uses radar to obtain images of the earth's surface. First to observe an ice free channel in the Arctic
CloudSat	Uses radar to observe clouds and precipitation from space; orbits in formation as part of a constellation of satellites (including Aqua, CALIPSO, PARASOL, and Aura).
Greenhouse Gases Observing Satellite (GOSAT)	The GOSAT was developed jointly by JAXA, Japan's Ministry of the Environment, and the National Institute for Environmental Studies (NIES) to observe the density of carbon dioxide, one of the gases causing the greenhouse effect.

(continued)

Satellite	Mission/Function
Clouds and the Earth's Radiant Energy System (CERES)	CERES is an instrument that has been used on several missions, including TRMM, Terra, and Aqua; it measures reflected solar radiation in several wavelength bands, which is used to measure the earth's total thermal radiation budget.
Moderate Resolution Imaging Spectroradiometer (MODIS)	MODIS is an instrument that has been used on several missions in the EOS program; it measures visible and infrared radiation to monitor vegetation, land surface cover, ocean chlorophyll, cloud and aerosol properties, fire occurrence, snow cover on the land, and sea ice cover on the oceans.
Terra	Terra is a multinational mission involving partnerships between NASA and the aerospace agencies of Canada and Japan; coordinates with the entire fleet of EOS spacecraft to study climate and environmental change.
Ice, Cloud, and Land Elevation Satellite (ICESat)	Measures elevation data needed to determine ice sheet mass balance; monitors cloud properties, especially for stratospheric clouds common over polar areas; collects topography and vegetation data around the globe, in addition to coverage over the Greenland and Antarctica ice sheets.
EOS Aura and Aqua	Daily global observations of earth's ozone layer, air quality. Monitors the earth's water cycle, including evaporation from the oceans, water vapor in the atmosphere, clouds, precipitation, soil moisture, sea ice, land ice, and snow cover on the land and ice. Studies radiative energy fluxes, aerosols, vegetation cover on the land, phytoplankton and dissolved organic matter in the oceans, and air, land, and water temperatures.
Tropical Rainfall Measuring Mission (TRMM)	Measures rainfall in the tropics and subtropics; tracks where heating and cooling associated with the rain are taking place in the atmosphere; tracks El Niños as they develop to provide input for global climate models.

APPENDIX D

Units of Measurement Applied to Climate Change

Prefixes Relevant to Climate Change

Prefix	Symbol	Factor	Power of 10	Common Name
Exa	E	1,000,000,000,000,000,000	10^{18}	1 quintillion
peta	P	1,000,000,000,000,000	10^{15}	1 quadrillion
tera	T	1,000,000,000,000	10^{12}	1 trillion
giga	G	1,000,000,000	10^{9}	1 billion
mega	M	1,000,000	10^{6}	1 million
kilo	k	1,000	10^{3}	1 thousand
centi	c	0.01	10^{-2}	1 hundredth
milli	m	0.001	10^{-3}	1 thousandth
micro	μ (Greek letter mu)	0.000001	10^{-6}	1 millionth
nano	n	0.000000001	10^{-9}	1 billionth

Temperature Units

Degrees Fahrenheit = 9/5 × degrees Centrigrade + 32
Degrees Centigrade = 5/9 × degrees Fahrenheit – 32

Measurement of Wavelength

Wavelength typically is measured in micrometers (μm) or nanometers (nm). A particular color red light has a wavelength of 0.65 μm or 650 nm. Either way, this light has a wavelength of 0.00000065 m.

Energy and Power

Energy is measured in joules. Electrical energy typically is measured in kilowatt-hours (kWh).

(A joule is about the energy it takes to lift an apple 1 m against the force of gravity.)

Power is measured in watts. A watt is consumption of energy at the rate of 1 joule per second.

The relationship between energy and power is: Power = energy/time.

Examples of the Use of Prefixes

1. The average power consumed per person in the United States is about 1.4 kW. This means that each person, *on average,* is continuously using 1400 W at all times.
2. The average energy used per person in the United States is 12 MWh per year. This means that each person, *on average,* consumes 12 million Wh of electrical power each year.
3. A nuclear power plant produces about 900 MW of power. This means that the power plant is continuously putting out 900 million W of power.
4. The Hoover Dam is rated at just over 2 GW. This means that the Hoover Dam is continuously putting out 2 billion W of power.
5. The average power consumption of the world is almost 2 TW. This means that the world is using, *on average,* 2 trillion W at any given time.
6. The power of a hurricane is 50–200 TW. This means that a hurricane has the power of 50–200 trillion W.
7. The power transported by the Gulf stream is 1.4 PW. This means that Gulf stream moves 1.4 quadrillion W.
8. The earth receives 174 PW of power from the sun. This means 174 quadrillion W.
9. The world used a total 490 EJ of energy in 2005. This means that the world's demand for energy is 490,000,000,000,000,000,000 J.

Examples of the Use of Prefixes

APPENDIX E

Selected Resources

Publications

1. Henson, Robert. *The Rough Guide to Climate Change.* London, Penguin Books, 2006.
2. Houghton, John. *Global Warming: The Complete Briefing,* 3rd ed. Cambridge, UK, Cambridge University Press, 2004.
3. Gore, Al. *An Inconvenient Truth*. New York, Rodale Books, 2006.
4. Linden, Eugene. *The Winds of Change*. New York, Simon and Schuster, 2006.
5. IPCC's 4th Assessment Report *Climate Change 2007, AR4: The Physical Basis and Impact, Adaptation and Vulnerability of the Fourth Assessment Report,* Intergovernmental Panel on Climate Change.

Web Sites

1. Intergovernmental Panel on Climate Change: www.ipcc.ch.
2. U.S. Environmental Protection Agency (EPA): www.epa.gov/climatechange.
3. National Oceanographic and Atmospheric Administration (NOAA): www.noaa.gov.
4. National Aeronautics and Space Administration (NASA): www.nasa.gov.

APPENDIX F

Summary of Key Climate Variables

Major Changes in the Earth's Climate

Variable	100 Years Ago	Today	Change
Average global temperature	13.7°C (56.5°F)	14.4°C (57.9°F)	+0.74°C (1.3°F)
Global carbon dioxide (CO_2) concentration	280 ppm	379 ppm	+35%
Global Methane (CH_4) concentration	715 ppb	1774 ppb	+248%
Global nitrous oxide (N_2O) concentration	270 ppb	319 ppb	+18%

Energy Balance

Solar constant (average value)	1368 W/m^2
Variation of the solar output over 11-year sun-spot cycle	±0.1%, or a range of 0.2%
Reduction in solar output during Maunder minimum (1645–1715)	0.15–0.3% less bright
Average power received on the earth's surface (taking into account its spherical shape)	342 W/m^2
Reflection from atmosphere, clouds, and earth	About 30%
Radiative forcing from greenhouse gases	+3.1 W/m^2
Radiative forcing from aerosols	–1.2 W/m^2
Radiative forcing from solar output variation	+0.12 W/m^2
Overall net radiative forcing (including additional contributors not listed here)	+1.6 W/m^2

Impact of Infrared Absorption

Theoretical temperature of the earth without atmosphere:	–19°C (–2.2°F)
Temperature of Venus	460°C (860°F)
Temperature of Mercury	170°C (338°F)

Persistence in the Atmosphere

Gas	Ability to Absorb Infrared Radiation (Compared with CO_2)	Lifetime in Atmosphere
Carbon dioxide (CO_2)	1	100 years
Methane (CH_4)	26×	12 years
Nitrous oxide (N_2O)	216×	115 years
Chlorinated fluorohydrocarbons (CFCs)	12,850–22,860×	45–1700 years

Rate of Global Temperature Increase

Time Period	Rate of Temperature Increase per Decade
Last 150 years	0.045°C (0.08°F) every 10 years
Last 100 years	0.074°C (0.13°F) every 10 years
Last 50 years	0.128°C (0.23°F) every 10 years
Last 25 years	0.177°C (0.32°F) every 10 years

Natural Cycles Affecting Incoming Solar Energy

Cycle	Range	Frequency
Sunspot cycle	Maximum and minimum	11 years (approx.)
Eccentricity of earth's orbit	Nearly circular to slightly less circular	100,000 years
Tilt of rotation axis	22.1–24.5 degrees	41,000 years
Precession (how direction of axis coincides with seasons)	360-degree rotation of axis	19,000–23,000 years

Natural Cycle	Key Characteristics
ENSO (El Niño/La Niña)	Recurs every 2–8 years; persists for 6–18 months

Predictions

Climate sensitivity (temperature if CO_2 doubles)	About 3°C, within range of 2–4.5°C (5.4°F, within range of 3.6–8.1°F)
Predicted temperature increase (for 2011–2030 compared with 1980–1999)	0.64–0.69°C (1.15–1.24°F)

Ocean Conditions

Average global sea level increase since last ice age	120 m (394 feet)
Average sea level increase since the 1800s	~1.7 mm/year (0.07 in/year)
Average current sea level increase (recent per TOPEX)	~3.0 mm/year (0.12 in/year)
Average ocean temperature increase	0.1°C (0.2°F) (from 1961–2003)
Average ocean pH increase	+0.1 (since 1750)
Arctic sea ice in summer	7.4% decrease
Arctic sea ice overall	2.7% decrease
Arctic ice-free in summer	By 2100

Potential for Contributing to Sea Level Increase (If Completely Melted)

Area of Melting	**Sea Level Increase**
Antarctica	56.6 m (186 ft)
Greenland	7.3 m (24 ft)
Mountain ice caps and glaciers	0.15–0.37 m (0.5–1.2 ft)
Permafrost	0.03–0.10 m (0.1–0.3 ft)
Arctic sea ice	0 m
Overall maximum potential sea level increase	64.4 m (211 ft)

APPENDIX G

Lingering Doubts and Concerns

Concern: How can we be sure the average global temperatures are really increasing?

Response: Temperature measurements collected from around the world show that the atmosphere and sea surface temperatures are increasing. These data were presented in Chapter 2 and show a statistically significant pattern of increasing temperature.

Concern: If the Earth is warming, how come some places are getting colder?

Response: Heat is constantly redistributed across the Earth after it is received from the sun. The Earth is not at a uniform temperature. Overall, the next change is in the direction of higher average global temperatures.

Concern: In the 1970s, scientists were worried about global dimming. They warned that the Earth was cooling. Why can't scientists make up their minds?

Response: At that time (as shown in Chapter 2), the Earth *was* in fact cooling slightly. Increasing air pollution (in the form of what is called *aerosols*) reflected some incoming solar energy. This caused slightly declining global temperatures. Today, aerosols still reduce the amount of solar heat that is received by the Earth, but the warming effect of the greenhouse gases is now clearly dominant. Chapter 6 details the relative impacts of the greenhouse gases, aerosols, and other components of the Earth's climate system on temperature.

Concern: Isn't the average temperature of the Earth increasing because of the simple fact that cities are hotter than the surrounding countryside?

Response: Precisely because of this (valid) concern, temperature measurement stations that might have been influenced by the heat island effect (that concentrates heat in urban areas) are excluded from global temperature averages. Urbanization and the structures of big cities are known to cause *locally* higher temperatures. However, organizations tracking the Earth's temperature are careful to *not* confuse the effect of urbanization with the global temperature increase. The heat island effect actually does contribute an identifiable but insignificant amount of added heat to the Earth. The heat island effect is discussed in Chapter 2.

Concern: Didn't satellite microwave sounding data actually disprove global warming?

Response: It is true that the first attempts to measure the Earth's temperature from space resulted in some initial confusion. The early satellites needed to resolve a change of a fraction of a degree each year from a distance of several hundred kilometers from the Earth's surface. As described in Chapter 2, the problem centered around the need to adjust data retrieved by the satellite by the precise orbital coordinates of the satellites. Now that this initial calibration has been performed, satellites are telling the same story as ground measurements, which is that the troposphere is warming slightly faster than the surface. The Earth is warming. The average global temperature is increasing and is warming at an increasing rate.

Concern: How do we know that global warming is not simply the result of natural rather than human drivers such as variability in solar intensity?

Response: Chapter 4 describes the natural cycles that can affect the Earth's temperature. Variations in orbital distance are accounted for by comparing changes from one year to the next. Changes in solar intensity are far too small to account for the climate changes that have been measured. Changes in the Earth's orbital conditions (the Milankovich cycles described in Chapter 4 occur over much too long a time interval to be considered a factor in the warming that has been identified in recent years. In addition, the temperature increase is consistent with the increase in greenhouse gas levels and sea level increase. Analytic techniques, such as isotope ratios and comparisons between the northern and southern hemisphere (as discussed in Chapters 5 and 6), point to a human fingerprint for global warming.

Concern: Are climate scientists being alarmist and exaggerating the impact of global warming?

Response: Scientists have identified a range of impacts of global warming. The Intergovernmental Panel on Climate Change (IPCC) has not issued a series of doomsday predictions but rather a carefully thought-out list of consequences based on data. (Consequences of global warming are addressed in detail in Chapter 7.) Climate models investigate a range of optimistic as well as "business as usual" scenarios. Sea level increases will be measurable but are not expected to flood coastal areas in the next few decades. Ice-free Arctic summers by the end of the century, increased droughts and floods from precipitation, reduction of the ice cover in Greenland, and intensified severe weather events are expected. Massive coastal flooding will not happen overnight and is not considered a done deal. However, continued buildup of greenhouse gases could irreversibly trigger that outcome. Disruption of the ocean currents that keep northern Europe more temperate is considered unlikely in the near future. Much of the Antarctica ice pack appears intact for the foreseeable future, although some melting is expected.

Concern: Aren't many of the climate changes beneficial?

Response: Some changes may be beneficial, such as reduced heating requirements in regions that now have severe winters. There may be extended growing seasons and more productive agricultural output in some areas. Cycles of drought and flood, water shortages, agricultural difficulties, and migration of disease-bearing insects are not beneficial changes. Nor are the massive coastal flooding and population displacement that could follow centuries of greenhouse gas buildup.

Concern: Won't correcting the problems of climate change create economic slowdown and stop commercial progress around the world?

Response: There are a number of economic opportunities in providing a solution to greenhouse gas emissions.

Concern: Is there anything we can really do to stop the progress of global warming?

Response: Yes. The technology to implement substantial greenhouse gas reductions is currently available. Substantial reduction in the burning of fossil fuels and the capture of the carbon dioxide from the fossil fuels (especially coal) that are burned are necessary to stabilize the level of greenhouse gases in the atmosphere. This is described in detail in Chapter 8.

APPENDIX H

Answers to Chapter Review Questions

Chap. 2	Chap. 3	Chap. 4	Chap. 5	Chap. 6	Chap. 7	Chap. 8
1. c	1. b	1. d	1. c	1. b	1. c	1. a
2. b	2. a	2. c	2. d	2. c	2. d	2. c
3. a	3. d	3. a	3. c	3. a	3. a	3. a
4. c	4. b	4. b	4. b	4. d	4. c	4. d
5. d	5. a	5. a	5. b	5. a	5. b	5. a
6. b	6. c	6. d	6. a	6. c	6. b	6. b
7. c	7. b	7. c	7. a	7. b	7. c	7. b
8. a	8. d	8. b	8. c	8. d	8. b	8. a
9. b	9. c	9. a	9. a	9. b	9. d	9. c
10. d	10. c	10. a	10. d	10. a	10. a	10. a
		11. c	11. b			11. b
		12. d	12. d			12. d

APPENDIX I

Answers to Final Exam

1. c	26. c	51. b
2. a	27. a	52. a
3. b	28. b	53. c
4. a	29. d	54. c
5. b	30. d	55. a
6. d	31. a	56. d
7. b	32. c	57. b
8. a	33. c	58. a
9. b	34. a	59. c
10. d	35. c	60. b
11. b	36. a	61. d
12. a	37. d	62. a
13. c	38. c	63. b
14. b	39. a	64. c
15. d	40. c	65. a
16. b	41. a	66. b
17. a	42. d	67. a
18. b	43. b	68. c
19. b	44. c	69. b
20. d	45. a	70. d
21. b	46. b	71. c
22. c	47. c	72. c
23. a	48. a	73. b
24. c	49. d	74. b
25. b	50. c	75. d

INDEX

F

G

H

I

J

K

L

M

N

O

P

R

S

T

U

V